THEOLOGY AND SCIENCE AT THE FRONTIERS OF KNOWLEDGE

NUMBER NINE

LOGIC AND AFFIRMATION
*Perspectives in
Mathematics and Theology*

THEOLOGY AND SCIENCE AT THE FRONTIERS OF KNOWLEDGE

1: T.F. Torrance, *Reality and Scientific Theology*.
2: H.P. Nebelsick, *Circles of God, Renaissance, Reformation and the Rise of Science*.
3: Iain Paul, *Science and Theology in Einstein's Perspective*.
4: Alexander Thomson, *Tradition and Authority in Science and Theology*.
5: R.G. Mitchell, *Einstein and Christ, A New Approach to the Defence of the Christian Religion*.
6: W.G. Pollard, *Transcendence and Providence, Reflections of a Physicist and Priest*.
7: Victor Fiddes, *Science and the Gospel*.
8: Carver Yu, *Being and Relation: A Theological Critique of Western Dualism and Individualism*.
9: John C. Puddefoot, *Logic and Affirmation — Perspectives in Mathematics and Theology*.
10: Walter Carvin, *Creation and Scientific Explanation*.

THEOLOGY AND SCIENCE AT THE FRONTIERS OF KNOWLEDGE

NUMBER NINE

GENERAL EDITOR – T.F. TORRANCE

LOGIC AND AFFIRMATION

*Perspectives in
Mathematics and Theology*

JOHN C. PUDDEFOOT

SCOTTISH ACADEMIC PRESS
EDINBURGH
1987

Published in association with the
Center of Theological Enquiry, Princeton
and
The Templeton Foundation
by
SCOTTISH ACADEMIC PRESS
33 Montgomery Street, Edinburgh EH7 5JX

First published 1987

ISBN 0 7073 0520 9

© John C. Puddefoot (1987)

All rights reserved. No part of this
publication may be reproduced, stored in
a retrieval system, or transmitted in any
form or by any means, electronic, mechanical,
photocopying, recording or otherwise, without
the prior permission of Scottish Academic Press Limited.

British Library Cataloguing in Publication Data

Puddefoot, John C.
Logic and affirmation: perspectives in
mathematics and theology.—(Theology
and science at the frontiers of knowledge; V.9).
1. Religion and science—1946–
I. Title
215   BL240.2

ISBN 0–7073–0520–9

Printed in Great Britain by
H Charlesworth & Co Ltd

To
Hilary, Rachel, Helen,
and Hannah

CONTENTS

General Foreword	ix
Prologue	xiii
Inheriting Doubt	1
Reconsidering Proof	41
Reintegrating Reason	88
Realising Truth	124
Living Understanding	155
Epilogue	197
Index	211

GENERAL FOREWORD

A vast shift in the perspective of human knowledge is taking place, as a unified view of the one created world presses for realisation in our understanding. The destructive dualisms and abstractions which have disintegrated form and fragmented culture are being replaced by unitary approaches to reality in which thought and experience are wedded together in every field of scientific inquiry and in every area of human life and culture. There now opens up a dynamic, open-structured universe, in which the human spirit is being liberated from its captivity in closed deterministic systems of cause and effect, and a correspondingly free and open-structured society is struggling to emerge.

The universe that is steadily being disclosed to our various sciences is found to be characterised throughout time and space by an ascending gradient of meaning in richer and higher forms of order. Instead of levels of existence and reality being explained reductionistically from below in materialistic and mechanistic terms, the lower levels are found to be explained in terms of higher, invisible, intangible levels of reality. In this perspective the divisive splits become healed, constructive syntheses emerge, being and doing become conjoined, an integration of form takes place in the sciences and the arts, the natural and the spiritual dimensions overlap, while knowledge of God and of his creation go hand in hand and bear constructively on one another.

We must now reckon with a revolutionary change in the generation of fundamental ideas. Today it is no longer philosophy but the physical and natural sciences which set the pace in human culture through their astonishing revelation of the rational structures that pervade and underly all created reality. At the same time, as our science presses its inquiries to the very boundaries of being, in macrophysical and microphysical dimensions alike, there is being brought to light a hidden traffic between theological and scientific ideas of the most far-reaching significance for both theology and science. It is in that situation where theology and science are found to have deep mutual relations, and increasingly cry out for each other, that our authors have been at work.

The different volumes in this series are intended to be geared into this fundamental change in the foundations of knowledge. They do not present "hack" accounts of scientific trends or theological fashions, but are intended to offer inter-disciplinary and creative interpretations which will themselves share in and carry forward the new synthesis transcending the gulf in popular understanding between faith and reason, religion and life, theology and science. Of special concern is the mutual modification and cross-fertilisation between natural and theological science, and the creative integration of all human thought and culture within the universe of space and time.

What is ultimately envisaged is a reconstruction of the very foundations of modern thought and culture, similar to that which took place in the early centuries of the Christian era, when the unitary outlook of Judaeo-Christian thought transformed that of the ancient world, and made possible the eventual rise of modern empirico-theoretic science. The various books in this series are written by scientists and by theologians, and by some who are both scientists and theologians. While they differ in training, outlook, religious persuasion, and nationality, they are all passionately committed to the struggle for a unified understanding of the one created universe and the healing of our split culture. Many difficult questions are explored and discussed, and the ground needs to be cleared of often deep-rooted misconceptions, but the results are designed to be presented without technical detail or complex argumentation, so that they can have their full measure of impact upon the contemporary world.

The Rev. John Charles Puddefoot, the author of this

GENERAL FOREWORD

volume, is a Master at Eton College, Windsor, and a part-time Chaplain. After reading mathematics at St Peter's College Oxford from 1971 to 1974, he started to train as an actuary, but soon switched to theology. He went up to Edinburgh in 1975 where he entered the Faculty of Divinity and after three years took a brilliant degree in Christian Dogmatics. He was ordained in the Church of England, serving first as a Curate in Darlington and then as Industrial Chaplain to Crawley in Sussex. He took up his present post at Eton in 1984 where he teaches mathematics and computer-studies. There he continues to develop his keen interest in the conceptual interrelations between theology and science. In this distinctive contribution to our series he wrestles with the fact that rationalism imposes upon us modes of thinking which threaten to cut us off from what is genuinely human. Mathematics offers an ideal system of logic which seems certain, while theological claims and methods seem arbitrary and loose in comparison. In this book John Puddefoot seeks to explore some of the illusions we have about mathematical certainty and to make room for a more personal approach to knowing and being in which all aspects of humanity can find expression.

Thomas F. Torrance

Warm thanks are due to the Rev. Robert T. Walker of Edinburgh for his cooperation in the revision of the proofs.

Edinburgh, June 1987

PROLOGUE

> THE Centipede was happy quite,
> Until the Toad in fun
> Said 'Pray which leg goes after which?'
> And worked her mind to such a pitch,
> She lay distracted in a ditch
> Considering how to run.
>
> Mrs Edmund Craster (Attrib.)
The Oxford Dictionary of Quotations.

While I was an undergraduate at Oxford I found myself in conversation with a fellow student reading the joint honours degree in mathematics and philosophy. He was giving up philosophy because he had found that it interfered with his mathematics: by thinking too much about what he was doing he had found that he could no longer do mathematics; by resenting the limitation of his mathematical powers he had lost the motivation to do philosophy. Certain kinds of question destroy our ability to focus upon what really matters. A mathematician relies upon automatic processes learned over many years to supply the insights which lead to solutions. To lose confidence is somehow to lose access to those processes, and to rely upon further sceptical self-consciousness to remedy the problem is to run the risk of lying distracted in a ditch for a very long time.

Transfer this anecdote from mathematics to religious faith and very little of the substance needs to be changed. Many theological students find the challenge to the foundations of their belief posed by critical analysis impossible to accommodate. The faith responsible for the vocations of many ordinands is slowly undermined by sceptical questioning, which leaves them unable to rediscover their original clarity of vision and purpose.

Philosophical and theological training has equated the process of hardening students to the realities of the world with initiating them into the rituals of institutionalised doubt: detached objectivity is regarded as the only respectable outcome of learning; uncertainty as the only proper manifestation of wisdom. While we can refer to doubts we can exonerate ourselves from any need for commitment; while we can refer to certain, but impersonal and mechanical explanations we can rid ourselves of responsibility. Therefore the domination of the principles of doubt and mechanism in our society can be attributed directly to our inadequacy as human beings. This is a vicious circle. Nevertheless, to point out that scepticism and mechanism dominate our intellectual outlook is merely to diagnose the disease from its symptoms. What we need is an analysis of causes and a cure. Even in mathematics confidence, rather than doubt, and humanity, rather than mechanism, are the touchstones of value and success.

SOURCES OF RENEWAL

We accept, actively or by default, some of the systems of authority we inherit with our culture. Such authority may be based upon individual relationships, as when we trust that someone tells us the truth; it may be based upon corporate experience, as when we trust that the community of scientists is not malevolently seeking to mislead us; it may be based upon cultural values, as when we accept the views about life, property, right and wrong encompassed by the laws of our society; it may be based upon trans-cultural experience, as when we share a common set of religious beliefs within a faith which embraces many otherwise different social and political ideologies. In some cases the authorities which we trust are able to persuade us to deny our own first-hand experiences. For example, science tells me that tables are not solid, but my senses tell me that they are; yet I accept the scientific claim because it arises from people and practices whose authority I respect. Sometimes we are not persuaded by the testimonies we usually trust, and find ourselves compelled to make personal decisions which involve dissent from the majority or established view. Sometimes we hear voices raised in protest at prevailing beliefs and values and where they may lead. Occasionally we may find ourselves called upon to affirm a weak tradition, to fight for a new belief or a new claim to knowledge which, like a man swimming against a strong current, has scarcely strength enough of its own to survive. Had none of our forebears been prepared to fight for such little things many of the springs which became streams, rivers and oceans of human experience would have been denied us. No more vital question can be conceived than what we should affirm and when and how we should affirm it. We cannot therefore eliminate our own authority, by which we determine when to believe, when to question, and when to deny.

Mathematics and theology, at least as popularly conceived, are as contrasting in their modes of authority as any two disciplines could be. In the first, incontrovertible impersonal proof based upon the strictest canons of logic seems to carry us forward inexorably; in the second, arguable personal conviction devoid of any satisfactory evidence or logically persuasive structure seems to leave us prey to every twist and turn of human fancy. Mathematics is regarded as authoritative because it appears to be characterised by clarity, certainty, objectivity and precision; theology is mistrusted because it is thought to be characterised by vagueness, indecisiveness, subjectivity and inaccuracy. In mathematics we imagine we have a very clear understanding of truth and proof, whereas in theology no such clarity seems attainable. We can weigh the obvious usefulness of mathematics as a framework for scientific advance against the apparent futility of theology, being drawn as a consequence to nod approvingly as professorial chairs in theology are frozen and dissolved as a result of cuts in university expenditure. Theology seems to be in eclipse, but mathematics remains the central tool of the sciences in their search for knowledge of the universe, a search still to a large extent motivated by the same desire for prediction and control which guided the ancients.

Yet not so long ago mathematics and theology were regarded as allies rather than enemies. We assume, and our experience seems to confirm, that the laws we derive at our desks and in our laboratories using mathematics accurately portray the character of the universe. But it is far from clear why this assumption should be justified; why, that is, the mathematical inventions of the human mind should confer upon us such awesome power, unless there is some further intrinsic harmony between the human mind and the structure of the universe, an assumption which many would regard as unwarranted and extravagant. Yet the ancients regarded the patterns which they saw in the heavens, the regularities of the movements of the celestial bodies, as evidence of the workings of the divine mind, and throughout history there have been those prepared to argue from the observed order of the universe to the existence of a divine creator responsible for its ordering principles. We like to believe that we have progressed beyond them, but despite the fact that as science has laid bare more and more of the secrets of the universe it has become necessary to make such arguments far more subtle, the conviction that behind the majestic complexity of the universe there must lie both a design and a purpose has never been extinguished, and the remarkable fact that the universe appears to be intelligible to the human mind remains to be explained.

Mathematics and the natural sciences have exercised a formative influence on our understanding of what it is to be rational, and by their manifest success they have persuaded us to adopt scientific rigour in assessing many claims to knowledge not directly connected with science or mathematics. But being fully human, fully personal, involves being moved by considerations which derive their force from more than logic and experimental demonstration. The question "what kinds of argument ought we to take seriously?" reaches beyond the confines of logic to redefine what we regard as an argument, and to free us to approach life at other than scientific levels. Whereas we seem to have a clear understanding of the nature of logic, we are far from an equally clear understanding of reason, and the claim that we have a perfectly clear understanding of what constitutes a proof in mathematics is far from true. Nevertheless we are perpetually tempted by the clarity of logic to make it synonymous with reason, and to disparage arguments which fall short of its canons. Our understanding of the kind of conceptual clarity central to science and mathematics has led to polarisations in theology between those who wish to share the status afforded to the sciences and to logical systems (and who consequently adopt versions of religious scientism and positivism), and those equally repelled by what they take to be the dehumanising logic of science (and who consequently espouse anti-scientific perspectives). It has also amplified the human tendency to migrate towards what appears to be simple, clear, tangible and certain.

Michael Polanyi, to whom a great deal in this book owes its inspiration, was prompted to turn from the practice of science to philosophy by his first-hand experience of science in Germany and Britain between the wars. He was appalled not simply by the political horrors it was made to serve, or by the narrow utilitarianism which guided it, but by the inability of more reasonable political philosophies to oppose such abuses of science effectively. He therefore set himself the task of providing a coherent philosophical basis for a new world-view to replace the liberal centre which had shown itself to be so inadequate. In the twentieth century we have seen a similar drift among religious people toward liberalism, fundamentalism, and sacramentalism, which derive their respective authorities from reason, text, and ritual. By drawing upon the contrasts and similarities between mathematics and theology as indicators of the assumptions of our age, I intend to argue for a *replacement centre* which is authoritative without being authoritarian, biblical without being fundamentalist, sacramental without being sacramentalist, distinctive without being irrelevant, clear without being dogmatic, and firm without being rigid. In other words, I hope to establish a position which converts the centre from being a mixture of mutually tolerant, but essentially disconnected and therefore ultimately powerless positions defined primarily in terms of what they *do not believe* and by an intellectual position based upon the Cartesian principle of the *suspension of belief*, into one in which the centre is united in the affirmation of a positive position based upon what it *does believe* and grounded in an intellectual position based upon *the suspension of doubt*.

Self-examination undermined the confidence that we had in our systems of belief and knowledge because it logically denied us a way back to faith. Having doubted, and having failed to put our doubts to rest, we had to remain unsure. Only in science, and particularly with the response to scepticism built around experiment and evidence, mathematics and mechanism, did it seem that we could allay our fears. Our lack of genuine self-confidence as living creatures came to contrast markedly with the confidence we had in things such as the sciences and the truths of mathematics. Man lost the courage to be a spiritual being and began to think of himself as a technician. Contrasts and similarities between theology and mathematics promise to shed light on this process.

Unless we are unable to see, we take sight for granted. We do not worry about how our eyes function, about how our optic nerves convey pulses to our brain, about how our brain interprets them to create the impression we call sight. Our perceptions of the world, and the things that we say about the world, rely on things that we take for granted which are just as delicate as the mechanisms of the body which we rely upon in sight. We see the world not just by virtue of physiological ability, but by virtue of a cultural framework which has been developed over centuries, that relies upon the work of countless others.

The development of our bodies from inanimate matter down through generations of slowly evolving kinds of creatures is paralleled by a cultural evolution no less wonderful and delicate by which we understand our world.

To do mathematics or theology we must have almost unconscious confidence in the underlying instruments by which they are done, and in order to live life we must have such confidence in the underlying instruments of living. If I attempt to do mathematics without adequate tools, however good my conceptual grasp, I may not arrive at the desired conclusion, be it a derivation of a formula or a proof of a theorem; and if I am in complete manipulative command of the tools available I may still fail to reach that conclusion by virtue of an inadequate conceptual grasp. In theology the same is true. We require a powerful synthesis of tools and imagination if we are to succeed in either, just as we require a healthy physiology and developed perceptual skills if we are to see correctly. One cannot simultaneously question the instruments, or the language and stuff of theology, and do theology. After Descartes we could never again be confident other than by an act of sheer will. That act of will, which was expressed consummately in Nietzsche's concept of the will to power, is arbitrary. It does not restore confidence by answering questions; it rides roughshod over them. And therefore the attitude "I believe this; this is right; do so as well", which Alasdair MacIntyre calls "Emotivism", arises not from extreme self-confidence, but from our inability to feel any sense of certainty. We cannot find a way back; we have been forced to camp somewhere else. But we have all chosen to camp in different places, there being no coherence to the positions which we adopt.

However happy we may seem, if we cannot find what it is to be human, we cannot do anything truly worthwhile with our lives. And each of us must make this discovery for himself. The program of discovering ultimately good, clear and secure *foundations* was bound to mislead us, and indeed has misled us, for it suggested that it is in secure *foundations* that the beginnings of understanding and the roots of progress are to be found. It can be seen especially clearly in mathematics and theology that the foundations do not guide the evolution of the structure, do not inspire, and do not renew, for they are by their nature fragmentary. Since Descartes each generation of philosophers has sought to remake the ruins of our certitude before feeling that we could again safely move forward. But the parts only make sense from the perspective of the whole. We therefore need to acquire again the ability to have visions and to dream dreams, to understand not by renouncing the ability to analyse, but by acquiring once again the ability, having performed our analysis, to return to commitment and faith.

What part can the Christian churches play in the renewal of this living vision? The majority of the ministers in the mainstream churches seem to be motivated by a sense of vocation grounded in one of the three positions I have mentioned. That is, they are drawn to the certainty and security of a precisely defined gospel or a precisely defined authority (in the shape of fundamentalism or authoritarianism), or to a position which is characterised by its opposition to such precision and authority. Members of the public are bewildered by the dispositions of the clergy, and starved of food which will enable them to grow in faith based upon belief in Jesus Christ as the Son of God. Even when they are well-disposed towards Christianity as they understand it they cannot respect a church which so manifestly lacks authority. They are invited to substitute *belief in the Bible* for belief in Jesus, or *belief in the rites and teachings of the church* for belief in Jesus, or *belief in the beauty of sceptical intellectual questioning* for belief in Jesus (or, in all three cases, to overlook or blur the differences between them). In the cross-fire between fundamentalism and authoritarianism (which demand that they believe and accept far too much without question), and liberal intellectualism (which demands that they question far too much without belief), the congregations of the churches are confused, and the unchurched lay public simply contemptuous. Evangelical churches and "high" churches often have far more respect for one another than they do for liberal intellectuals. They recognise that it is more important to *believe something* than to *question everything*. The liberal intellectuals, on the other hand, argue that it is more important to *question something* than to *believe everything*. Is there no position which involves believing enough to keep our questions in check, and questioning enough to keep our beliefs in check?

Fundamentalism and sacramentalism are concrete instances of what I later call "formalism" or "absolutism". I am not opposed to evangelism, and I am not suggesting that we abandon the sacramental practices of the Church; I am suggesting that wherever we find that the words of the Bible or the rites of the churches are being treated as sufficient in themselves we encounter invalid forms of depersonalisation aimed to remove our responsibility for what we believe and do by locating it in "objectivity" based upon an impersonal "authority". In particular I have in mind, wherever the Polanyian idea of perfectionism leading to inversion occurs, the extraordinary rise of anti-science, anti-intellectual movements here in Britain and in the United States, which now force us to take seriously ideas such as Creationism which were dead even ten years ago. These movements draw their strength from the dehumanising scepticism of so much science and philosophy, which forces ordinary people into irrational positions out of a vain hope that they will thereby preserve their humanity.

These problems must be treated at two different levels. First, questions concerning the status of the Christian message and the basis of its proclamation must be addressed as they bear upon central assumptions about the nature of human existence in order that we can avoid the impossible choice between irrational faith and cynical detachment. My task in the present volume is to make some headway towards these goals. Second, the question of ministry must be addressed from this new perspective, because in the absence of a clearly articulated theology capable of withstanding the assaults of sceptic and extremist the churches will be tempted to define their roles in secular terms. I hope to explore the second issue in a later volume.

I have generally avoided technical discussions of the philosophy of mathematics, where issues of the greatest complexity arise, because the series of which this volume forms a part is intended for interested non-specialists. To enter into such discussions would necessitate burdening the reader with a great deal of mathematics, which I have been at pains throughout to avoid. By showing that mathematics can be redeemed from the charge that it is merely a matter of manipulating formulae I shall provide evidence for the thesis that all knowledge, reason, and truth must be brought within a reshaped understanding centred upon relationships of faithfulness to our world, our neighbours, our selves, and our God. The present volume is therefore only a first step upon a long road, offered as a gesture of hope for the future, and in confidence that we have one, rather as Jeremiah purchased the potter's field all those years ago.

INHERITING DOUBT

FEAR OF FAILURE

THE contrasts between mathematics and theology with which this work began echo prevalent values of our age in distinctions drawn and separations made between reason and emotion. By preferring dispassionateness, impartiality and objectivity we adopt a value-system which places impersonal categories above personal ones because we distrust our emotions. But our emotions are real and important to us, and so we contrive to channel them into culturally acceptable modes of expression which do not interfere with reason.

Our suspicion of the emotions stems from such things as our personal and cultural experience of their fallibility, destructive capability, and fickleness. We seek to control them using reason and self-discipline, social custom and tradition. But might not each of these four mechanisms itself be untrustworthy and systematically misleading? How can we be sure that our culture does not contain serious flaws, that our reason is as reliable as we would wish, given that most of our doing, knowing and believing must of necessity be undertaken without rigorous examination? Considerations such as these led Descartes to propound his Method of Doubt as a prerequisite of his project of Pure Enquiry.

> I had long ago noticed that, in matters relating to conduct, one needs sometimes to follow, just as if they were absolutely indubitable, opinions one knows to be very unsure ...; but as I wanted to concentrate solely on the search for truth, I thought I ought to do just the opposite, and reject as being absolutely false everything in which I could suppose the slightest reason for doubt, in order to see if there did not remain after that anything in my belief which was entirely indubitable. So, because our senses sometimes play us false, I decided to suppose that there was nothing

at all which was such as they cause us to imagine it; and because there are men who make mistakes in reasoning, even with the simplest geometrical matters ..., judging that I was as liable to error as anyone else, I rejected as being false all the reasonings I had hitherto accepted as proofs.

Descartes: *Discourse on Method 4.*

Descartes strips away social custom, sensation, and habitual reason in rapid succession to reduce what he knows to the barest minimum, which he concludes to be the famous *cogito ergo sum*, "I think, therefore I am". The one thing in which Descartes cannot be mistaken is that he is *res cogitans*, a thinking thing.

It is not my purpose to discuss Descartes' philosophy in detail, but to observe the sense in which this whole program is motivated by his awareness of the possibility of error, and therefore by the prospect of failure. The sheer unreliability of appearances, intuitions, passions and emotions presents us with the likelihood that we might at some stage fail by making wrong assumptions or drawing false conclusions. Descartes' remedy is extreme caution, the adoption of a method which applies the severest possible tests to any claims to certainty and allows through only such things as appear indubitable. The passage cited is one of the most famous sources for the subsequent histories of scepticism and individualism: scepticism because it suspends belief in hitherto accepted truths; individualism because it rejects social custom in favour of the certainty of the introspecting *ego*. Henceforward, if we follow Descartes' method, we can trust ourselves only as thinking things, not as bearers of sensation and emotion; some of those features which are irreducibly human have therefore already been set aside in the quest for certainty and infallibility.

Those aspects of human reason which are most precise are also most easily communicated, especially using mathematical forms. Since the emphasis upon the individual requires clarity of *thought*, and clarity of thought tends to depend upon clarity of expression, this form of individualism leads us to value ideas which are communicable clearly, distinctly and unambiguously using formal language. Individualism inspires a corporatism devoid of non-formal content, that is, devoid of content which cannot be set out solely in terms of formal expression. Formalism is born.

On the other hand, the project of scepticism cannot allow that anything of genuine importance lies beyond the subject-matter of pure enquiry, for to do so would be to restrict its scope to an unacceptable degree. Therefore from the practice of scepticism there arises the need for rationalism, a philosophy which requires all things to be rationally intelligible without remainder, and which regards all things which are not apparently intelligible in such a way as regrettable blemishes which we should ideally remove or exile to some harmless province of human affairs. Mathematics itself can supply many examples where intuition, so often a mathematician's friend, leads him astray, and many of the antinomies (paradoxes) which trouble philosophers of mathematics arise from apparently innocent notions (such as that of a set), and make them mistrustful of other intuitively clear ideas.

This account outlines the way in which from an acute awareness of our fallibility the nexus of philosophies consisting of scepticism, rationalism, individualism and formalism emerges. But with their emergence there is an inevitable erosion of the irreducibility of the human, so that fear of failure and love of certainty threaten to erase the sense of humanity which gives rise to them. This is an example of what Michael Polanyi calls *perfectionism* giving rise to *inversion*, the destruction of the values which inspire the pursuit of perfection by the means of its pursuit. In the process of trying to produce a perfectly moral society we may be tempted to legislate in ways which will destroy not only the society we would protect, but the morality we would preserve, by taking absolute power in order to impose absolute obedience. In the process of trying to produce a perfectly safe and certain theory of knowledge we may surround our knowing and enquiring with such restrictions that we eliminate from what can then be called knowledge the very insights and values which motivated the enquiry: we create a vision of the world from which man is absent; all that is left is "atoms and the void". Or, as Steven Weinberg puts it, "The more the universe seems comprehensible, the more it also seems pointless".

THE OBSTRUCTION OF HUMANITY

The task of the Church is to further the purposes of God, who created the world to provide an environment in which his creatures could grow into sons and daughters. Those forces and fashions which subvert that program will therefore form the focus of a Christian critique of the world. If the analysis to be presented in these pages is correct, the principal deficiency of our culture is that it prevents us becoming persons by disparaging those qualities in man which are characteristically human. In part, such deficiencies are an occupational hazard of the scientific enterprise, where a proper concern to understand what man is made of can easily become confused with the view that man is nothing but what he is made of. It is as irreducible a feature of human nature that we have visions, beliefs and dreams as it is that we have fears, doubts, and nightmares. To destroy the possibility of the dream for fear of the possibility of the nightmare is "to make a wilderness and call it 'peace'". But with dreams and visions, beliefs and feelings comes the possibility of error, and with error comes the possibility of failure. Therefore, if dreaming is irreducibly human, the possibility of failure is inescapably part of the human condition and we must learn to live with it and benefit from it.

Consider an analogy. Anyone who plays a musical instrument knows how important exercises such as scales are. But to become a pianist one must play more than scales and arpeggios; the only way to learn to play pieces is to play pieces. If someone suggested that since pieces are nothing but notes, and in playing scales we play all the notes on our instrument, therefore we need play no pieces, we would regard him as mad, as having completely missed the point. But exactly the same is true of human existence: to become fully human we must practise those skills which are uniquely human, such as believing and dreaming and loving, and learn to accommodate the less welcome features of existence which are also intrinsically human such as dying, suffering, losing, and failing. In a cultural climate which implicitly doubts the status of such human skills there is a strong possibility that they will not be practised, and therefore will be lost; and where we also contrive to deny the limitations of our finitude there is every danger that we will never learn to rise above it. Stated in its strongest form this means that *we do not know what it is to be fully human.* I do not say that we "no longer" know what it is to be fully human, for that would suggest that our practices introduce blemishes into what is essentially established. On the contrary, I make the much stronger claim that these habits of mind *prevent* us from proceeding along the path to full humanity, to attainment of the status of God's sons and daughters, from the gift of existence to realisation of our essence.

A dilemma of the greatest practical importance now emerges. The individualism which arises from mistrust of social practice, whether in the form of tradition or contemporary authority, requires the very highest standards of each individual life if it is not to reduce to social anarchy. Each of us must be able to decide for himself what to do on every conceivable occasion. (It must be said that Descartes did not regard this as the aim of his system; there is ambiguity in his writing as to whether he envisages his successors accepting things on his authority, or working everything out anew for themselves.) Scepticism and individualism involve rejecting authority and tradition; therefore they must suppose that all questions have short, finite answers which can quickly be regenerated to replace tradition and fill the vacuum of authority scepticism has created. They involve optimism about the ability of each individual to recreate from scratch, without presuppositions, values, morals, practices and attitudes which have taken centuries to evolve, or to generate new systems of values, morals, principles and attitudes of equal or greater legitimacy otherwise. We not only can but must make up our own minds on all such questions if some hidden or overt form of authority or binding tradition is not to be reintroduced.

The view of *mind* implicit in this philosophy is undeniably individualistic, as it must be if it is to be true to Cartesianism; but can a philosophy of mind be individualistic given what we now know about the structure of language? We should not be misled by the obvious separation between *brains* to infer an equally obvious separation between *minds*, since the stuff of minds (thoughts, beliefs, desires, intentions, and ideas) are both culturally conditioned and culturally dependent. Without the body corporate I would be "mindless". Therefore the notion that I can make up my own mind from presuppositionless beginnings is philosophically very curious. But as soon as I allow the merest doubt about my ability to do this the problem emerges of how I am to decide which values, morals, principles and attitudes to generate for myself, and which I am to accept on the basis of some kind of authority or tradition, however disguised.

In this dilemma we can see one misleading feature of the mathematical ideal. Mathematics, although it deals with open-ended structures and recognises as-yet-unsolved problems, nevertheless deals with *finite* questions which have beginnings and endings. A mathematician knows when he has solved a problem, and so do all his contemporaries, just as they know when it remains unsolved (although the advent of computer proofs has raised an area of controversy even here; cf. the dispute over whether the four-colour theorem has or has not been proved). Moreover the answer is, to a large degree, final. There are almost *no* problems of this sort in philosophy, or in everyday life. There are certainly problems which acquire finite solutions ("would you like tea or coffee?"), but nobody supposes that this is a matter of necessity rather than expediency.

The time-dependence of everydayness pinpoints a further deficiency in mathematics as an ideal for human reason and certainty: mathematics is a-temporal; everydayness is not. Mathematical problems are non-finite in just the sense that we are finite, and finite in just the sense that we are non-finite. That accounts for both the power and the problem of mathematics.

It might seem that, if these conjectures are true, the solution to the problems which have ensued from scepticism is straightforward. All we need do is to eschew the method of doubt and we will discover what it is to be fully human. That is not the case, for the influence of scepticism is not most severe at the level of the self-conscious practice of doubt, but where the structure of doubt is built into our civilisation, and passed on from generation to generation.

CULTURE, NURTURE AND CHOICE

New-born babies inherit certain kinds of abilities genetically, but we expect their culture to supply them with language and other skills necessary for survival, the importance of which for our development into mature adults makes the attitudes and values prevalent in our culture a matter of general concern. But each culture is different, and no culture remains the same for long. Ideas valued highly in one generation can easily be forgotten by all but historians in the next. Suppose, therefore, that a valuable idea ceased to be generally available, so that even historians of ideas could scarcely reconstruct its former appeal and importance? An illustration may make this clearer.

When we go shopping we usually have a list of things that we need. Most of those things will be readily available, but some may not be on the shelves. However, some substitute product may catch our eye, or some complete novelty, which tempts us to spend our money on that instead. Besides these three cases (things we need that are available, things we need that are not, and things that we suddenly discover a need for), there may also be things we do not think that we need, and which are not on the shelves to awaken a need, things which would provide great nourishment but which are out of fashion. We are limited, in other words, by what we regard as *possible*, which is governed by what our culture teaches us (where "culture" is being used in a very wide sense, of course, in these days of world-wide communication). Of the things which we know to be possible, some will be in fashion and some not because of prejudices and conventions concerning which of the things that are available are also desirable. Where ideas are concerned we have to take account of implicit rather than explicit propaganda. There are very few active missionaries for the cause of a scientific world-view, or the methods of scepticism, but criteria drawn from science and scepticism permeate society. They are for sale on the shelves in the shop of ideas, and they are what people know to be available. There is no need for us to be conscious that we are selecting from a subset of the total human experience, or even be familiar with or sympathetic to the philosophies of Descartes and his successors in order to adopt them. In just the same way we can support capitalism without having read Adam Smith, or Marxism without having read Marx or Lenin. We are guided by, and make use of ideas, long before we have cause to examine the adequacy of their foundations, as the history of mathematics over the last hundred years shows.

SYMPTOMS AND SOURCES OF CHANGE

One of the first signs that something is wrong with a tradition is that its members begin to experience unease and to express dissatisfaction with life. Negatively, they may begin to turn aside from their own tradition and seek something in another, apparently more exotic culture (as we have seen in the attraction of India to many young people); they may develop illnesses of the mind or body; they may behave in irrational ways by turning to such things as astrology and occult practices out of boredom with their lives and with the stock of ideas available from more conservative sources; they may experience insecurity, and form allegiances with any group offering straightforward answers to their problems. Positively, there may arise the beginnings of an alternative world-view from the observations of sensitive men and women able to detect the signs of the end of an era.

The possibility of articulating a renewing world-view depends upon the existence of a vocabulary (a formal system) and a conceptual world (a non-formal system) sufficiently clearly distinguished from the current ones to catalyse change. One such possibility is that we should again speak of God out of a tradition reaching back to Abraham, Isaac and Jacob. This is one mode of discourse, whose concepts and vocabulary it is the responsibility of the Church to safeguard in order that speech about God should continue to play a part in the conversation of the world. Nevertheless, neither the possibility nor the means of preservation of that mode of discourse should be taken for granted. The possibility is eroded wherever the concept of God is blurred and clouded, diluted by various interest-groups who wish to make it in some sense more palatable to themselves or to others; the means of preservation is undermined wherever the Church falls back upon emphasis on formal expressions of belief (dogmas) without a corresponding non-formal spiritual revival.

COMMUNITIES OF ENQUIRY

We depend upon a community of like-minded people if we are to keep our bearings, benefit from a common mode of discourse, and communicate genuine insights which will uplift, guide, and sustain further discovery. The notion of a group of like-minded people sharing a vision of some aspect of the world by which they correct one another Polanyi called a *conviviality*, a community of enquiry which regulates itself by constantly referring its development backwards and forwards to the object of its enquiry. Polanyi had in mind in particular the community of scientific enquiry. Where questions of God and Jesus are concerned, the conviviality is of course the Church.

The conviviality of the scientific community shapes its corporate criteria of truth, proof, value, and scientific advance. The usage of a scientist must either conform to current practice (demonstrate that the words used exhibit correct understanding of the concepts they name), or be of sufficient forcefulness to persuade the scientific community to revise existing usage in favour of new concepts. This is then regarded as a scientific advance. The point is that it is the community of *scientists* which acts as the arbiter of advance in this respect, not the mass of non-scientific humanity. The scientific community collectively sets its own standards and agrees to be bound by them. In this sense it is dogmatic, in that to refuse to comply with prevailing standards is either to attempt to change them, and risk exclusion and ridicule if one fails, or to place oneself voluntarily outside the broad scientific community.

In this respect the churches must take responsibility for the divergent views expressed by those who hold high office within them. For a bishop to profess himself unconvinced of the truth of the Virgin Birth, and to dispute the historicity of the Empty Tomb, and still be adjudged worthy of his episcopacy is in effect for the church concerned to have accepted that degree of divergence from orthodoxy, for by analogy with the scientific community such an expression of dissent must either place the dissentient outside the sphere of the church's deliberations (he is untheological in precisely the sense that a scientist similarly placed would be deemed unscientific), or be accepted as one whose views one would be willing to be convinced by (as one whose dissent promises to play a significant and constructive role in deepening and enriching the churches' teaching and understanding).

Scientists and mathematicians are fortunate to have virtually universally agreed standards by which new theorems and hypotheses are tested, and to be dealing with a subject which admits of a conceptual clarity and formalisation of sufficient precision and rigour to ensure that any result can, in principle, be checked by anyone else suitably qualified. This is not in general the case in other disciplines, and most of the humanities are troubled from time to time by theories which divide the academic and lay public. This is nowhere more true than in philosophy and theology, where schools of thought raise bitter antagonisms which can descend to the level of personal abuse in attempts to discredit one or other theory. Emotion cannot easily be excluded from human affairs.

That this is the case makes it all the more lamentable that the churches lack sufficient unanimity and doctrinal clarity to speak authoritatively. The problem is not that truth will suffer; truth can be left to look after itself. But our access to truth and to clarity of language and thought cannot be separated because, as Frege taught us, we rely upon the clarity of our language and its associated concepts for the clarity of our thoughts. Confusion over our usage of key terms such as "God" and "Resurrection" threatens us with religious anarchy, and with a debasement of religious language which threatens our very capacity to think or speak intelligibly about such matters at all. For most people the traditional sense of the word "God" has become inaccessible — a precise and disturbing example of the kind of erosion of meaning that occurs when the community entrusted with and concerned for the establishment and preservation of standards fails in its duty to speak with anything approaching a coherent voice. The Church has a duty to sort out what tolerances it will allow. No scientist would be permitted to use a word such as "electron" in a completely novel way without being called to account by his peers.

Another reason for this erosion of meaning arises from

consumer pressure. Religion has become a consumer-led industry supplying food which lacks real nourishment. Churches are under pressure to supply religion understood as some form of meditative and spiritual activity, apropos nothing in particular, intended to make us better, more whole people, or a more integrated, patriotic nation. There is nothing wrong with the intention to become more whole, but we do well to remind ourselves of Karl Barth's insight that religion can easily become unbelief, a supreme example of human sin in its attempt to achieve self-sanctification, which may be translated as the attempt to make ourselves feel at one with ourselves, the world, and our version of "God", without it costing us too much, and preferably in such a way that we enjoy the process as much as possible. But by cutting Christianity down to a more popular size we relinquish its capacity to stretch and extend people. On the food analogy, it may taste pleasant, but it has negligible nutritional value.

DISTINCTIVE PROCLAMATION

Sociology of Religion seeks to analyse what the churches *ought* to do on the basis of what people *want* the churches to do. This idea is so old that Socrates anticipated it, as in the following famous passage from Plato's *Republic*:

> Suppose a man was in charge of a large and powerful animal, and made a study of its moods and wants; he would learn when to approach and handle it, when and why it was especially savage or gentle, what the different noises it made meant, and what tone of voice to use to soothe or annoy it. All this he might learn by long experience and familiarity, and then call it a science, and reduce it to a system and set up to teach it. But he would not really know which of the creature's tastes was admirable or shameful, good or bad, right or wrong; he would simply use the terms on the basis of its reactions, calling what pleased it good, what annoyed it bad. He would have no rational account to give of them, but would call the inevitable demands of the animal's nature right and admirable, remaining quite blind to the real nature of and difference between inevitability and goodness, and quite unable to tell anyone else what it was.
>
> (VI:493b, c.)

So it is with those who seek to make the sociology of religion the basis of theology. But the attraction of such an approach is strong where empirical methods are held in high esteem, as they are in our science-dominated culture. The "is" absorbs the "ought": this is how things are, therefore this is how they ought to be. If the churches are seduced by this methodology they will pay a high price for the pseudo-status which results: renunciation of any sense of moral priority (i.e. any sense of a constraining authority which might attempt to improve the "is" by reference to the "ought"); assumption of a dispassionate value-free stance ("we merely report the facts, we do not comment upon them; we do not make *value-judgements*"); implicit acknowledgement of "numerical might is moral right", a version of the Democratic Fallacy (see below).

The debate about the churches and politics, more especially about the Church of England and political pronouncements, can only be resolved when the churches realise and acknowledge that they speak the language of the world with an understanding that at least attempts to be divine. To use words merely as the world uses them serves only to add fuel to an already over-stoked fire. Politics becomes increasingly clearly a battle between individualism and corporatism, the political philosophy of the entrepreneur based upon human greed and selfishness, versus the political philosophy of the masses based equally upon human greed and selfishness, the difference being that in the former case we are encouraged to make our own fortune, and in the latter we are encouraged to rely upon someone else to do so. For the churches to contribute to this debate without pointing out the inadequacy of these alternatives only serves to reinforce the mistaken attitudes that have made us a divided nation and a demoralised church. The Church is concerned about people, and politics arises wherever there are people; therefore the Church is concerned in politics. But it should comment *as Church* only insofar as it has something distinctively Christian to say.

Lesslie Newbigin drew attention to some of these

problems in his widely read essay *The Other Side of 1984*. Rumour has it that he was driven to write it out of exasperation with discussions at the British Council of Churches, which he felt relied upon an uncritical adoption of post-Enlightenment values and assumptions. In sixty pages he summarises a program which he hopes will revitalise the churches for a new dialogue with modern culture from a standpoint which is not hopelessly relativised by that culture. The need for the churches to think through what their ultimate intentions are is nowhere more apparent than when, in making political pronouncements, they betray their lack of a clear vision of the society they are working to bring about. The articulation of such a vision is a theological task, for it must be generated in the context of the kind of double focus which incarnational theology requires, namely a view of history which sees God working in the world without doing disservice either to the world (true man) or to God (true God). It is unfortunate that in all-too-many of our pronouncements our convictions about the purposes of God for the world seem of only peripheral importance.

The access of the Church to an alternative viewpoint will always depend upon its total life of prayer, where "prayer" is to be understood in the broad sense of its total corporate orientation towards the being of God. To fail to speak distinctively is to demonstrate the paucity of our theology, and to confirm the suspicions of the unchurched that we lack access to any source of inspiration other than the world. Talk of other-worldly inspiration is unfashionable, for we are persuaded that what we cannot explain cannot occur; we have put proof before facticity. This is one reason why Barth's theology is so often disregarded: we cannot in general bring ourselves to acknowledge the possibility that the "wholly other" might reveal himself because we cannot imagine how he might manage to do so. No-one presumes (quite rightly) to answer the question of how it is possible for theology to begin from God (rather than from man) and we are inclined to draw the conclusion that it cannot begin from God, that the ground of theology cannot be the self-revelation of God, but must rather be the self-understanding of men. This split between the knower and the known is typical of the dualism that has entered into Western thought since Descartes as a result of scepticism, and of the mistaken principle that the only things which can be countenanced as real are the things that human beings can understand.

Mathematics depends upon conceptual clarity. To drive mathematical formulae in the right direction we must have a grasp of the non-formal principles which govern them. Knowing statements of principle *as statements* will not enable us to apply them to novel problems; they must be understood as well as known. Moreover, it is only by trying to apply them to novel circumstances that we discover whether we understand them as fully and correctly as we hitherto believed. In much the same way we discover the adequacy of contemporary theology by its response to novel problems. The cry constantly goes up that the Bible offers no guidance on nuclear weapons, abortion and embryo research as specific issues. But neither do mathematical principles embody all their spheres of application. Rather, my grasp of those principles is put to the test by the adequacy of my response to an unforeseen situation. On this basis our theological understanding stands under judgement, if not condemnation. It is not that the churches have nothing to say, but that what they say sounds too often like a thinly-veiled espousal of one or other party-political philosophy.

AXIOMATICS AND DOGMATICS

How we explain the failure of a system to respond adequately to new circumstances depends upon our view of the relationship between a system and its formal presuppositions.

In his interesting book *Axiomatics and Dogmatics*, J. R. Carnes sets out to show that dogmatics in theology is isomorphic to the formal aspects of science, i.e. mathematics. This involves drawing strong analogies between the two disciplines which prove extremely illuminating, but are not without their dangers. My principal objection to his interpretation is that it involves making mathematics (and therefore, implicitly, theology) a science of formula-manipulation, dealing with axioms as givens and building theorems from them. This understanding of mathematics is very misleading because it treats axioms, whether in mathematics or theology, as givens, as if axioms themselves have no history. They become "take it or leave it" data, the foundations of structures developed from them; and dogmas, on Carnes' interpretation, have the same role in theology. This locates the roots of systems in quite the wrong place: an axiom set is not the foundation of a system, but the product of generations of mathematical enquiry as it has eventually been formalised or *axiomatised*. This point is beautifully expressed in a textbook on abstract algebra I used when an undergraduate at Oxford:

> We should like to stress that these algebraic systems and the axioms which define them must have a certain naturality about them. They must come from the experience of looking at many examples; they should be rich in meaningful results. One does not just sit down, list a few axioms, and then proceed to study the system so described. This, admittedly, is done by some, but most mathematicians would dismiss these attempts as poor mathematics. The systems chosen for study are chosen because particular cases of these structures have appeared time and time again, because someone finally noted that these special cases were indeed special instances of a general phenomenon, because one notices analogies between two highly disparate mathematical objects and so is led to search for the root of these analogies.
>
> Herstein, Topics in Algebra, p. 25f.

Telling the story of axiomatic mathematics without the story of its development loses its heuristic ground, and is, I suspect, responsible for a large proportion of the mystique and despair surrounding the way mathematics is taught. Let us call this *the sin of retrospective refinement*, of presenting results which have been developed by the sweat of one's brow, and after wasting reams of paper, in neat ways which hide their true path to birth. The student sees only the final product, and can be forgiven for being overawed and depressed by it on account of his awareness of his own scribblings and fumblings. How can he ever hope to be as good a mathematician as his teacher, who produces such neat and short proofs in comparison? Little does he suspect that the teacher very often has had to fill his own waste-paper basket with just as many sheets of scrap paper before arriving at these neat solutions. In terms of the collective efforts of mathematicians, the sin is compounded by text-books which present the final analytic solution without any adequate account of the heuristics, the fumblings and gropings which are an essential part of any discovery.

I am labouring this point because a similar problem bedevils theology, and leads people to regard the dogmas of the Church as givens which spring from nowhere which they must swallow mindlessly. The New Testament is not a "first time onto paper" account of the experience of Jesus; it is material that had been predigested for years before reaching its final form. The statements of the creeds were not "first time" solutions to Christological problems; they represent the results of agonising debates conducted over centuries. Yet if we miss out this historical detail we condemn generations of Christians to the same kind of indigestion to which students of mathematics are condemned: to understand the axioms of abstract mathematics, or the dogmas of the Church, it is necessary to see them not as foundations, but as *distillations* of a corporate act of enquiry. Analytic theology can no more survive without heuristics than can analytic mathematics; more to the point, neither is there any reason why it should. But the process whereby we regenerate the non-formal understanding with the help of the formal axioms, dogmas, proofs and theorems is far from straightforward.

Carnes begins his book with the following wonderful quotation from Werner Heisenberg:

> One of the most important features of the development and the analysis of modern physics is the experience that the concepts of natural language, vaguely defined as they are, seem to be more stable in the expansion of knowledge than the precise terms of scientific language ... This is in fact not surprising since the concepts of natural language are formed by the immediate connection with reality; they represent reality ... On the other hand, the scientific concepts are idealizations; they are derived from experience obtained by refined experimental tools, and are precisely defined through axioms and definitions ... But through the process of idealization and precise definition, the immediate contact with reality is lost.
>
> Heisenberg, *Physics and Philosophy*, p. 200.

Heisenberg is saying that in the process of abstraction and axiomatisation (retrospective refinement) *we lose contact with reality*. But in our everyday language, for all its looseness and imprecision, we are conscious that the opposite is the case: despite our inability to account for the purchase ordinary language affords us upon reality, we are confident that in using the vocabulary of tables and chairs we are in touch with what is real (the senses in which we can also trap ourselves with this self-evidence will concern us in the chapter on truth). We do not stumble over sentences, hesitate to construct coherent arguments, struggle to find suitable expressions for commonplace ideas. In more technical matters such as philosophy and theology it is necessary to be far more careful about terminology, but that is because our grasp of the concepts involved is less clear, and we must try to avoid confusing ourselves. This feature of language would only be true of the mathematical sciences as understood by a tiny number of great men, that we use it for the most part with complete confidence and conviction.

Oddly, Carnes does not make use of this implication, but uses Heisenberg's words to substantiate his argument that theology "is no more empowered to legislate for religion than science is to create worlds in competition with that of ordinary experience" (p. 2). Theology, that is, "consists of theories about religion and the objects of religious experience". It must account for religious experience on the basis of axioms supplied by religious practices (such as the Bible, the Creeds, and the doctrines of the Church). But the result of this is that whenever dogmas no longer refer to the data of religious experience (i.e. when the experiences to which they refer no longer occur) theology becomes "freighted with existentially irrelevant terms" (p. 84). "A theology is 'useful', religiously speaking, if it explains, helps us to understand, religion and the religious" (p. 85). Here the dangers of Carnes' "isomorphism" between mathematics and theology become clear, for in mathematics we can generate a multitude of different systems merely by changing the axioms, and the analogy forces upon us the acknowledgement that in theology we can do the same. The adequacy of "a theology" (note the indefinite article) is thus to be judged by its ability to explain religious experience, which leads inevitably to a pragmatic relativisation of truth (all Carnes' caveats to the contrary acknowledged): "we are entitled to judge that a set of religious beliefs is true, in a perfectly full-blooded sense of 'true', if those beliefs hold together consistently and if they enable us to make sense of our lives and our world" (p. 120). In other words, as he goes on to say in the last two pages of the book, the concept of truth reduces to something internal to a system: Marxists and Christians both hold true beliefs because they are consistent; unfortunately they are also incompatible, but that possibility is an inevitable and insurmountable feature of axiomatics. These small points of difference between my own position and Carnes' add up to a major difference over the nature of truth and the obligations it imposes upon us which precludes acceptance of his solution to the problem of incompatibility. This will concern us again later.

Heisenberg seems to me to be saying very clearly that in the precision of our formalisation we lose contact with reality, and therefore that (contra Carnes) the need for a controlling vision of the reality to which our formal systems refer is increased the more precise and clear they become. If theology is to be true to God the *Datum* of Revelation is utterly essential to the theologian, not redundant; and it is especially necessary where we decide (as we must) which religious experiences to include as the axioms of our Christian theology.

FORMALISM

One argument running through this book is an attack on *formalism*, on ideologies and practices which replace the categories of the *living*, which are uniquely organismic and, more particularly, human, with their formal expression. The currency of formalism is impersonal and transportable knowing and being, of which one example is mechanism. But formalism arises wherever we treat abstractions from concrete instances as sufficient in themselves.

How appropriate is the *abstraction* we have already seen to be so essential to science and mathematics, whether as an escape from reality or as the means whereby we invent tools to impose our will upon the world? Is the price actually or potentially too high that we pay for this power? Do we, in choosing the grand generalisation and clarity of pure thought, lose touch with the particularity and individuality of the real world in which we live, as Heisenberg suggests? Is there a hidden trade between perfection and personhood? Is the legacy of Plato, the dogma that in discovering the most general and most pure we also discover the most valuable and the most true, actually *false* and *destructive*? In T. S. Eliot's words,

> Where is the Life we have lost in living?
> Where is the wisdom we have lost in knowledge?
> Where is the knowledge we have lost in information?
> *Choruses from "The Rock" I.*

We take for granted the kinds of *thought experiments* which entertain *unreal assumptions* for the sake of conceptual clarity – "what if ...?" questions — but do those experiments divert us from the most real path by seducing us into pursuing an unrealisable ideal? Is the fact that we have developed technologies more powerful than the wisdom we have developed to control them actually an *inevitable* consequence of this division of the living from the hypothetical? In other words, once we accepted (as we did almost without consideration) the desirability of pure abstract thought, did we perhaps condemn ourselves to an evolutionary phase which could not do otherwise than lead us to this divorce of the possible and the desirable?

The scale of the problem we face is apparent from the obvious sense in which anything approaching an affirmative answer to these questions necessarily involves a large-scale rejection of both the technological aspects of the rise of science *as it has actually occurred* (but not necessarily in principle), as well as the principal legacy of Greek philosophy, the duality of pure thought and concrete existence, of theory and practice.

The modern history of this development arises of course from Descartes. Not, as is often supposed, because of the consequences of scepticism or the Method of Doubt themselves, but because of the *principle* which is presupposed in their exercise. Descartes is falsely accused by many of advocating a philosophical system which if severely applied would render all life impossible. But Descartes was as aware of this problem as any of his successors (cf. Bernard Williams, *Descartes* p. 61ff). No, the problem arises *before* the consequences of scepticism are noticed (serious as they are); it arises from and with the assumption that asking questions self-evidently divorced from real conditions is a good thing to do. The question must be posed whether a system of thought or a method which self-confessedly defies any attempt to practise it in everyday life is not condemned from the outset to lead to disaster when pursued by beings incapable of living with its answers, whether, that is, it is not bound to lead to the "two worlds" of science and everydayness which Carnes so wishes to avoid (op. cit. p. 1).

I am grateful to T. F. Torrance for drawing my attention to the thought of James Clerk Maxwell in this context concerning the desirability of a sense of *embodiment* in mathematics. Clerk Maxwell's researches into the electromagnetic field forced upon him an awareness of the trade-off between generality and specificity in mathematics, that in disembodying mathematics by abstraction we lose our purchase upon concrete particular situations. This relates to the point Heisenberg is making in the quotation above, and is loosely reflected in the Uncertainty Principle itself. What is not perhaps as clear is the trade between *power* and *personality*, between the *generality* conferred upon us by abstraction and the *value* we experience when in touch with uniqueness. For it is paradoxical that the individualism which has flowed from scepticism has given rise not to a process of what the mediaevals called *individuation* (the generation of unique individuals), but to the apparently contradictory process of *institutionalisation*. By being forced back upon ourselves as ultimate arbiters of truth and falsehood, and being all-too-conscious in most respects of our inadequacy for such a responsibility, we become prey to adoption of *default* ideals as transmitted by our culture: the perfectionism which advocates each individual deciding upon everything for himself is inverted by the processes whereby individuals renege on this responsibility and adopt the ruling values and ideologies of their day.

Consider an illustration from an apparently unrelated field. During five years of marriage counselling I noticed again and again that the problems which were bedevilling the couples I saw arose from *phantasies* about possibility and reality, that is from unreal expectations about the nature of life and the possibilities and capabilities of human beings borrowed from the institutional phantasies of our era. Husbands and wives who failed to be sexual athletes, romantic lovers, successful in business, popular in society, or perfect parents, were resented by their spouses as *failures*, and their actions and inactions were interpreted as symptoms of love grown cold. These considerations lend support to the view that the high level of divorce in our society springs more from our excessively *high* expectations of marriage than from our indifference to it. Our expectations cannot be fulfilled because nothing can fulfil expectations based upon a distorted perception of the nature of the world. We have failed to accommodate the power of theory to the necessities of practice. (A vital ingredient in the diet of any commentator on our times consists of the value-systems and supposed possibilities portrayed in advertisements on independent television.)

If the remedy to difficulties such as those in broken marriages is a greater sense of the truth about reality, the objection will be raised that reality is something from which most of us feel we need to escape from time to time. The real world is a tough place full of viruses, earthquakes and other man-made and natural disasters; the last thing most of us feel we need is a larger dose of it. But the question then arises whether this picture of reality is itself real. Is our relationship with reality as distorted as the relationships between marriage partners; do we expect too much from it, and torture ourselves with unfulfillable desires? There is little doubt that to torture oneself with regrets is only to heap coals upon one's own head. But neither is the cynic's beatitude the way out: "blessed are they that expect nothing; they shall not be disappointed". Rather, "blessed are they whose expectations are real; they shall be fulfilled". Our indignation at international disaster and personal misfortune is fuelled by some dimly-formed picture of a world in which such things do not occur, perhaps a remnant of a vision of the Garden of Eden. Yet once we stop running away, nursing our phantasies and resenting reality, and turn and face the world, it is extraordinary how quickly even the most severe setbacks and disasters can be accommodated.

Even, or perhaps especially in religion, there is this element of escapism, of generating a false picture of the world filled with false hopes. But in Christianity this is particularly remarkable, for the central reality of the Christian Gospel is the Cross, which represents the quintessence of reality at its harshest, most unjust and most cruel. By romanticising the Cross we remove the core of Christian realism, that despite even this possibility, life is good. The churches then become the homes of those who cannot face reality, who need some prop upon which to lean, or some drug upon which to grow dependent, whereas they should be the homes of those who, in Charles Davis' words, "can absorb it in doses which would unhinge the ordinary person" (in *Theology and the University*, ed. John Coulson).

All this serves to confirm my suggestion that we do not know what it is to be fully human, and it leads to the further contention that we have lost sight of the true nature of the Christian Gospel. Alasdair Macintyre provides an analysis in his book *After Virtue* of a similar situation regarding morality.

AFTER VIRTUE

MacIntyre begins by inviting us to imagine a world in which, as a result of some catastrophe, science is dead. The books and formulae remain, but none of the understanding of science that gave them life. In attempting to make sense of this diverse body of material, "What would appear to be rival and competing premises for which no further argument could be given would abound" (p.2). He goes on, "The hypothesis which I wish to advance is that in the actual world which we inhabit the language of morality is in the same state of grave disorder as the language of natural science in the imaginary world which I described." Thus our moral discourse lacks a base and reduces to *emotivism*: "I approve of this; do so as well." Why? The Greek synthesis (which reached its culmination in Aristotle's *Nicomachean Ethics*) was divided into a choice between authority and reason, where failure of reason led to appeal to authority, which was therefore devoid of a rational base. Morality had either to be justified rationally or by appeal to authority. As the authority crumbled (in the shape of Church, Bible and State), so the emphasis fell upon reason, but reason required also the source and destination consensus of "man as he is" and "man as he would be if he realised his true end"; it was in fact the means of transforming the first state into the second. However, the notion of a true human end supposes the notion of an "end" to be legitimate, and just this was rejected in the overthrow of Aristotelianism by modern science, especially its teleological explanations and purposive vocabulary. Thus reason found itself with "man as he is" without any agreement on "man as he would be if he realised his true end". But devoid of a destination, reason cannot show us how to pass beyond "man as he is"; consequently we have stayed there.

The choice between authority and reason remains apparent in the distinction I have drawn between absolutism and liberalism, between arbitrary authority in the shape of fundamentalism or sacramentalism, and aimless reasonableness in the shape of the liberal intellectual. The choice seems to be exclusively between static and circular modes of being. The kind of reintegration of authority and reason which Carnes seems to envisage, on the model of axioms and logic, leads only to a static and arbitrary formalism by using mathematics as a model for the reintegration of authority and reason by analogy with the assumption of an axiom system (authority) and its extenuation by mathematical logic (reason). Then theology, as contrasted with religion, deals with the axioms of the faith (texts, credal statements, doctrines) in a similar way in order to provide further theological propositions.

The vacuum generated by the failure both of authority and of reason, of hierarchy and teleology as MacIntyre expresses it, was filled by the invention of the autonomous individual whose end is to realise his true self as individual, and whose morality is to be decided upon by himself in pursuance of that goal. An early, tame, but recognisable prototype of the *Übermensch* is born. And so, necessarily, are emotivism and democracy.

Emotivism must follow from individualism because *my*

experience of what is good *for me* is the only basis of my value system. In trying to communicate with you, I therefore rely upon what seems good to me. Hence, "I approve of this; do so as well". Any possibility of another conception, say in terms of corporate well-being as such, is lost, because corporate well-being is intelligible to individualism only in terms of its ensuing benefits for individuals. We find ourselves devoid of the conceptual tools (linguistic terms and metaphors) necessary if we are to conceive of existence in any other way (because our culture has nurtured us on a diet of pure individualism). Even the death of Jesus is interpreted as the supreme self-sacrifice of an individual on behalf of other individuals: *his* life for *my* salvation.

Democratic ideals must follow from individualism because we are forced into corporate compromises by the human situation. Since the unit is the individual, corporate decisions can only fairly be taken on the basis of aggregates of individual preferences. Thus democracy is a form of collective emotivism: "enough of us approve of this; do so as well". The criterion of rightness and wrongness becomes a democratic majority. No attempt is made to justify actions otherwise than by reference to such a principle. In some respects it can be argued that no attempt is made to justify actions at all; might (democratic superiority) has become synonymous with right (the will of the majority). The difficulty with this, of course, is that the individual units are voting not according to their perceptions of truth or corporate benefit, but in pursuance of their own self-realisation. Democracy feeds on the food of individual self-interest, greed and envy, and defines that which optimises the satisfaction of individual self-interest, greed and envy as "the good".

> They hide in the throng and call numbers to their aid.
> Tumult.
>
> Pascal, *Pensées*, Penguin Edition, p. 208.

Some of the forms and consequences of the Fallacy of Democracy are as follows: (a) if the number of people voting for the truth of position A is equal (or even roughly equal) to the number of people voting for the falsehood of position A (not-A), then whether A is true or not is a matter of opinion, since the evidence cancels out; (b) because people disagree about what is true, everything and nothing is true; (c) because there is disagreement about truth there is no truth; (d) if something can be doubted it cannot be true; (e) because there are people who disagree with you, you have *no right* to be convinced of the universal truth of your position (i.e. the falsehood of competing positions); (f) truth is decided by weight of numbers.

In these forms the Democratic Fallacy gives rise to a sense of outrage that anyone, being aware of the possible falsehood of a position, and knowing that many sane men and women actually regard it as false, should self-consciously and deliberately set aside those doubts, and affirm its truth and the falsehood of other positions. Polanyi demonstrated the inconsistency of positions which pretend that it is improper or unnecessary ever to make choices which involve rejection of competing perspectives in the absence of totally convincing counter-evidence by referring us to world-views conflicting with our own. We do not reject the views of primitive cultures about gods and demons, a flat earth and witch-doctoring, for example, on the basis simply of *evidence*, and we do not allow their views to undermine our confidence in our own understanding simply because they are divergent. We reject them because we regard them as *false*, and even as *absurd*. If ten per-cent of the population of the world voted that it was flat, would I hold that it was round with only ninety per-cent conviction? Of course not; the idea is ridiculous. But the selection of which positions to reject is not, as it would be for the emotivist, *arbitrary*, a matter of mere whim.

Loyal to the dogmas of individualism we all like to think that we make up our own minds on most issues. In fact we accept parental, societal and third-party opinions as our guides in the majority of cases. We are each in love, not with the reality of individual responsibility, but with an abstract idea of it. We like to think that we are free individuals because we resent the thought that we are any man's servant, cipher or slave. The ideals of the American Declaration of Independence are our "self-evident" ideals, including the right to "Life, Liberty and the pursuit of Happiness". We resent the notion of authority, of a "higher power", of being responsible to anyone or anything but ourselves. In particular, we resent the idea of God.

However, the achievements of the last four hundred years were based upon the power of abstraction and generalisation. Gradually, by virtue of cultural transmission and dissemination, the currency of idealisation overwhelmed our capacity to analyse reality as we experience it in our daily lives. A process we might call the *institutionalisation of thought* robbed us of our autonomy almost before we had learned that we were free from dogmatic authority. With the growth of mass communications, huge populations came to be fed with the thoughts of comparatively few analysts. The paradox emerged of societies composed of people who in theory are autonomous, self-directing agents, capable of making up their own minds on every conceivable subject, while the educational and social practices of their culture render them incapable of doing so. Our populations are therefore vulnerable, for example, to the ridiculous suggestion that they can buy a small share of an ideal home and an ideal, loving relationship of trust and family harmony with their children, just by deciding to buy a particular brand of washing-up liquid.

SCIENCE AND SCIENTISM

The achievements of the mathematical sciences represent one of the greatest flowerings of the human spirit. Their principal benefits, including medical advances, provision of time and labour-saving devices, and improvements in communication and our general standards of living, rightly establish the authority of the sciences in our culture. But these achievements lead some to suppose that *all* human knowledge should conform to scientific ideals, and that knowledge which fails in this respect should not be regarded as knowledge at all. This view is called *scientism*.

The attraction of such a view derives from the *impersonal* nature of the sciences as popularly understood, the *respect* for the sciences in our culture, and the *insecurity* we feel in our individualism. In particular we avoid uncertainty and personal investment in our knowing and doing by espousing objective criteria based upon a mixture of overt facts and social customs. To believe what scientific standards credit as facts, and behave in what society regards as a normal way is to make oneself inconspicuous and to minimise or eliminate risk. But once again the pattern of perfection and inversion emerges. Individualism, based upon an extravagant over-estimation of the capacities of individuals to decide for themselves what morals, values, principles and attitudes to adopt, places so much pressure on us all that it destroys our confidence even in such abilities as we genuinely have. Consequently it leads not to the extension of personal competence, but to its erosion in the cult of the impersonal; not to increased demand for human contact and relationship, but to increased demand for artificial and inauthentic modes of being. This difference can be described schematically in terms of *active* and *passive* lifestyles, between doing things for ourselves and allowing other people to do them for us, between participating in and contributing to society out of the resources of our own uniqueness, and acting as an external spectator who observes but does not join in. The difference between playing tennis and watching tennis is harmless enough, but carried to extremes the same kind of difference can drastically limit the part we are able to play in our community. The success of science in prolonging life and saving time throws into sharp relief the question what all the life and time so saved is for? Have we reduced the working-week from seventy to thirty-five hours so that people can watch thirty-five hours of television, or extended life-expectancy by ten years so that we can be bored for ten years longer? If we are to use the time and life saved by science profitably we have to rediscover the dimensions of the human and personal which, paradoxically, the scientific ideal as reflected in scientism serves only to erode. The prospect of more intimate personal contact with other people frightens us because we neglect nurture of the personal skills required.

Our desire for impersonal theories of knowledge, for neutral ground, for some completely objective basis upon which to decide between truth and falsehood, is an example of what Sartre calls "bad faith". It is an attempt to circumvent what it is to be human in the human situation by sloughing off the responsibility we have to decide as a matter of ultimate personal commitment and choice, guided by our participation in a culture and tradition, participant as we are in a community of enquiry, and as such on the basis of received authority, what is true and good and just and honourable, and consequently what we are to do with our lives.

DOUBT

One of the epistemological problems Christians face is that Cartesian scepticism has been so influential in Western thought that we find it is almost impossible to understand how, once a religious doubt has been voiced, it could ever legitimately be silenced without recourse to some form of dogmatism. The possibility that we might be wrong introduces an inescapable division between knower and known. To overlook that possibility is to abandon the scepticism which has been so productive in science by making us aware of the possibility that we might be mistaken.

In the chapter on commitment in *Personal Knowledge*, Polanyi insists that we must preserve the tension between dwelling in our knowledge fully as a subsidiary means of attaining focal knowledge of something beyond it, and reserving the right (which we would frequently exercise) to withdraw from that indwelling in order to examine those subsidiaries. Otherwise we find ourselves back in the instrumentalist-versus-realist problem in linguistics and axiomatics which Carnes mentions in the first chapter of his book. The connection between our language and reality, our knowing and what we know is achieved not by analysing our language or knowing at a meta-level (as though there were some neutral ground upon which we could stand devoid of misconceptions from which to examine the correspondence between our language or formalisation and reality), but by dwelling in it in such a way that it conveys to us more than any mere symbolism could otherwise be expected to do. A formal system can only properly be applied or manipulated when (a) its sense or meaning is understood by the person using it, and (b) it is allowed to refer him to the reality beyond itself.

THE TRANSPARENCY OF THE FORMAL

As I have argued in the next chapter, it is the function of a proof to help us to expand the concepts to which it relates. It is a characteristic of any formal system, and especially of language, that we find ourselves incapable of giving a fully adequate account of what we mean by our formal terms or words. But by elaborating upon the bare skeleton of a sentence or theoretical statement we come to understand what lies behind and around it. These features of formal systems lie behind the modern philosophy of language claim that we know the meaning by the use, and discern whether or not someone else understands by paying attention to the way he speaks, writes or generally uses words.

> The understanding, then, does not reside in the words themselves, nor even in the appropriateness of the whole sequence of words and sentences. It lies, rather, in the fact that an *understanding* speaker

can *do things* with the words and sentences he utters (or thinks in his head) *besides* just utter them.

Putnam, "Language and Philosophy"
in *Philosophical Papers II* p. 4.

If we demand that someone gives a precise analytical account of his concepts we are unlikely to receive more than a shallow answer which does not go far towards explaining to us his undoubted but nevertheless remarkable ability to use and recognise the use of the word in quite novel and unexpected circumstances. In other words, he knows a chair when he sees one and would immediately spot an inappropriate use of the word. This commonplace, but nonetheless striking ability is only an example in everyday speech of what we aim for in all human study, namely that our concepts should become instinctive, transparent to the laborious processes we endure to acquire them. It is my concern to re-establish the grounds for similar confidence in our lives as personal human beings with beliefs, hopes, visions, and dreams, but that is possible only if we can bring ourselves to set aside certain doubts deliberately. (A mathematician who stops believing that he can solve problems stops being able to solve them.)

The process of learning a musical instrument provides a further striking example of this acquired transparency. The laboured, even painful childhood process of drilling the fingers to play scales and arpeggios, trills and turns, must eventually give way to an automatic skill which is at the command of the performer beneath the conscious level, for the focus of his attention as a concert pianist or violinist can never be upon the mechanics of playing, but must remain unfalteringly upon the shape and meaning of the music. It is in this capacity to conceptualise entire pieces of music, and not in the prodigious feats of dexterity that are prerequisites for performing them, that we detect genuine artistic genius. The first is a necessary, but not sufficient condition for the second, whereas the demands of the total conceptualisation preclude concentration upon the actions of fingers. The technique must first be acquired and then equally necessarily forgotten as far as conscious attention is concerned.

Problems we face in all study parallel this passage from pain through competence to exhilaration, for whereas a grasp of some elementary facts is necessary if anything is to be achieved, a mere grasp of those facts is no guarantee of either competence or talent in a subject. Formalism, for reasons which will become clearer towards the end of the next chapter, mistakes the necessity of formal expression for sufficiency. Scepticism makes doubt a procedure of such rigour that it forbids recourse to any route back from it.

> Any particular commitment may be reconsidered ... after which, having satisfied his doubts, the reflecting person would recommit himself ... But he would find the return blocked if, having realised that this movement involves an act of his own judgment, he denied justification to it by reason of its personal character.
>
> Polanyi, *Personal Knowledge*, pp. 303ff.

Knowing therefore involves creative acts of judgment and understanding.

FACTS AND CREATIVITY

I am reminded of an example given by Berman of the way children have been known to use their home computers to access databases, and to respond to requests for essays on, for example, the Reformation, by producing print-outs containing every reference to the word "Reformation" available. Some are unable to perceive that this is not, and can never be what is required, but it is nevertheless instructive to ask oneself why, precisely, this is not acceptable, and what that can tell us about the nature of human knowledge and scholarship.

Human beings are engaged in interpreting their world all the time. Language itself is a system of interpretation, and does not afford us an undistorted view of the world. The children in the database story were attempting to present their teachers with "raw facts", and we may conjecture that their estimate of the value of their homework was that the more raw facts they unearthed the greater the knowledge they had acquired. We might even anticipate that they believed that "raw" facts, uninterpreted and therefore undistorted facts, were of more value than facts as digested and regurgitated by their adolescent brains, being more "objective". But there are no "raw facts", for all facts are of necessity expressed in some language, all language must depend upon an extremely complex conceptual system to be intelligible, and the colour of that conceptual system in turn shapes both their expression and interpretation. Moreover, contrary to the supposition that raw facts are the stuff of scholarship, what we value in a scholar is his skill at building up a new conceptual scheme by which to account for the "facts" which shows how they relate to other aspects of experience. Understanding does not manifest itself in the ability to manipulate pieces of information mechanically, but in the capacity to use pieces of information in novel and yet persuasive ways. Far from feeling that we are in touch with reality when presented with a set of facts, we feel that reality draws near when we can conceive of it in a way which makes sense of the facts and yet which also goes beyond a mere restatement of them.

This is another way of saying that all true acts of understanding are *creative*. We commonly speak only of *authorship* and creativity together: Shakespeare, Mozart, Leonardo and Einstein are in their respective ways "creative"; we deny creativity in those who watch, hear, appreciate or understand their work; but this is a prejudice. The child who uses a database to write an essay on the Reformation without first absorbing the information it unearths is an extreme example of human life in a culture which does not realise that creativity is as necessary to appreciate a work of art or science as it is to produce it. The spectre of objectivity lies behind our reluctance to acknowledge that this is the case. As soon as we allow that we contribute in some way to our appreciation of the world out of the well of our own experience the suspicion is aroused that we have lost objectivity and with it truth; that the pure milk of raw facts has been poisoned with the bile of personality. Beauty and profundity are conceived as a one-way traffic from author to devotee, rather as science is conceived as a one-way traffic from the universe to the scientific mind, whereas the devotee in appreciating beauty gives something to the work of art as surely as the author does by his labour.

Polanyi expressed this point of view by insisting that all true knowledge is personal; it is something in which the knower is personally involved. Without that involvement knowledge is impossible. Theories of knowledge which pretend that it is unnecessary and undesirable hide the truth behind another instance of bad faith by pretending that we are not responsible for what we know, and that we play no part in our knowing. Such "objectivity" claims that knowledge forces itself upon the knower in such a way that it is simply not open to him to deny it; therefore he is not responsible for it. This is the great Enlightenment lie, a lie which Kant delineated in his doctrine of the autonomy of man, that no man can deflect responsibility for his own beliefs and actions by pointing to his need to obey a higher authority because in the end it is his own decision to obey that authority; and which finds expression in the legal principle that a soldier or policeman cannot divest himself of responsibility for his actions merely by indicating that he was acting under orders.

Unlike Kant, however, Polanyi never suggested that the mind constitutes reality, or that we create truth by our acts of commitment; he argued that we must each carry our share of the collective responsibility of any peer-group for the truths it affirms, especially when they subsequently prove to be false. Only those prone to the Fallacy of Democracy mistake such responsibility for constitutive action. The epistemology arising from Cartesianism insists on the polarity of fact and opinion, facts being those assertions which survive the razor of scepticism, opinions everything else. With Empiricism the pattern changes, but the result remains the same: facts are assertions proved by experiment, impersonal and absolute; everything else is mere opinion. Polanyi insists that this antithesis is the source of our problem, not its solution. Personal knowledge arises from *persuasion*, which will usually involve empirical, theoretical and other components, but in the last analysis it depends upon my conviction that one thing is the case and not another, a conviction arrived at in the context of my sense of ultimate responsibility for my beliefs. Nothing can obviate the need for such personal participation in knowing, but fortunately we can usually rely upon a community which shares our enquiry and enables us to try out our ideas with like-minded searchers after the truth. But (and this is the part of the argument that causes all the trouble) *acceptance of one view involves rejecting competing views*; in fact, I am under as much obligation to reject as I am to affirm, for it must be the case that what I believe to be true I believe to be true *for everyone*, even if my insistence upon its universality proves to be inconvenient or painful for me.

The concepts which shape my usage of language must certainly be shared to some degree with others for us to be able to communicate, and my peer-group certainly has the right to challenge my understanding of certain words when my usage differs from the norm; but in the end my usage can only be governed by my concepts, and therefore what I hold to be truths are my responsibility. To the extent that I share a collective misunderstanding (perhaps built into the grammar and vocabulary of my mother-tongue, or the cultural heritage of my nation) that individual responsibility is lessened; but to the extent that I further distort it or reduce it, it is my own. In this way, as a fully active and responsible member of society, I make my contribution to the tradition and benefit from the contributions of others. Only so is it possible to rely upon the received wisdom of the "tribe", and only on the basis of such a society (a conviviality) is the confidence we have in even mathematics and science justified.

The reconstruction of an affirmatory theology which

this book seeks involves rejecting the choice between a view of human rationality which is limited to the criteria of the natural sciences (scientism), and a freedom of belief which countenances such things as astrology and witchcraft as proper human interests. The customary compromise which accommodates these extremes is the liberal doctrine of toleration whose motto might be the remark attributed to Voltaire, "I disapprove of what you say, but I will defend to the death your right to say it". Carnes envisages a collection of mutually exclusive but equally "true" systems along similar lines. In this he is true to the philosophy of mathematics espoused by Tarski, but at a price, as Putnam points out:

> Tarskian semantics gives no explanation of the meanings of "true" and "false" when they are used to compare and criticize different theories, if meaning is really theory-dependent.
>
> op. cit. p.x.

All these problems arise because having entered into the procedures of scepticism we can conceive of no way out of them again. They present us with what might be called an epistemological black hole which sucks everything and everyone in, but allows nothing it swallows to leave again, so great is its pull. By denying the personal qualities which enable us in our everyday lives to make necessary choices, they prevent us from acknowledging the possibility that certain avenues of thought might be true in ways which imply that others are false, in particular that transformation of our understanding in favour of an orientation towards the world which is richer and more fulfilling for which St Paul used the expression "the mind of Christ".

TELEOLOGY

It is characteristic of science that since the Enlightenment it has rejected recourse to final causes to effect explanations of phenomena discovered in the world. This rejection was partly justified as a protest against Aristotelian animism which conceived of the universe as an organism subject to and guided by the same goals and desires we all recognise and experience, a philosophy which served to suppress what came to be regarded as "real" explanations. But science paid the price of losing the vocabulary of purposive action in this rejection of Aristotle, so that when it came to apply the methods of science to human beings it attempted to explain behaviour which can only legitimately be described in terms of goals and purposes without reference to goals and purposes. The result could only be a growing tension between the claims of science and the experience of people in their everyday lives. The analysis that I have given of the unacceptable alternative between narrow rationalism and broad irrationalism describes one consequence of this tension. The methods of science, empirical and theoretical, have assumed too dominant a role in our assessment of what it is to be human and rational, and threaten to prevent us from realising our true nature by circumscribing what we feel we can believe and affirm. Yet it is those things which we feel we can believe and affirm which provide us with our ambitions and goals, not the scientific or other means whereby we achieve them. Therefore it is necessary, for the health and survival of the human race, for us to have a fund of visions, goals and beliefs about ourselves and the world which will supply worthwhile and attainable ends. In the absence of such a fund we will, as I have said, merely dissipate our energies in chasing unreal and escapist pursuits which will herald and partly bring about the decline of our civilisation.

The need for a synthesis of the natural and human sciences can be seen from our need for levels of description which are other than quantitative, in which a limited range of purposive categories is admissible (categories which, for reasons alluded to above, are generally regarded as outside the domain of the sciences), and from our need for a depth of explanation which proceeds beyond the equations and formulae of mathematics. The overthrow of final causes in post-Enlightenment science in favour of supposedly more "real" explanations has served us well for four centuries, but now, in an age increasingly bereft of direction, the blind mechanisms of chance offer no hope or guidance, and new categories of explanation are required. The Christian Faith offers us just such an integrative explanatory system, capable of supplying us with new goals which will alter our perception of our condition and the opportunities we have to change it for the better, because it combines an insistence upon the intelligibility of the universe with an equal emphasis upon the value of human categories of thought and action. But if we follow Carnes in making theology a descriptive, passive, theoretical discipline we will have renounced what qualifies it to speak distinctively and authoritatively, namely its bearing upon the Truth.

CHEAP REPENTANCE

Rationalist criticism has prompted the churches to repent of the faith of their fathers and to seek comfort and popularity by adopting a more "rational" (that is, sceptical) attitude to the stories of the Bible and the doctrines of the Early Church. As I read Newbigin's essay the awful implications of this hasty repentance impressed themselves on me by comparison with the book of Job. Job was told by his "comforters" that his misfortune must have arisen from some sin he had committed, and advised that he need only repent to be restored to health and wealth. Job refused because he knew himself to be blameless before the Law, and because he would not repent unless convinced of his sin. Unlike Job the churches have sought to ease their sufferings by listening too readily to latter-day comforters like Eliphaz, Bildad, Zophar and Elihu; they have too readily repented of their "dogmatism" and "irrationalism", of their lack of scientific evidence, of their dependence upon distant history, and of their lack of irrefutable proofs for their claims. But the words which crushed Job's resistance were not uttered by any human voice: "where were you when I laid the foundations of the world?" The churches would do well to remind themselves of their source. In that court, and in the presence of that Judge, and confronted by that reason, they would learn that they had uttered words they did not understand, and been charged and invested with a burden too wonderful for them to bear. Then they would learn to see with their eyes what they had only heard with their ears, and justly despise themselves, and repent in dust and ashes.

RECONSIDERING PROOF

IT is commonly supposed that mathematics is made what it is by virtue of the availability of rigorous proofs to establish its truth, and that the notions of proof and truth it employs must therefore be completely clear. These assumptions concern us in the present work because the mathematical ideal casts a shadow over all other kinds of argument and assertion: over against a mathematical "prove it!" there always stands a sceptical "but you can't be sure!" Ever since Anselm advanced the Ontological Argument in his *Proslogion*, and Aquinas classified the *five ways* in his *Summa Theologica*, there have been many attempts to frame a satisfactory proof of God's existence along mathematical lines. All these attempts fail, fascinating as they are, and valuable as they have been in prompting clarification of theological and philosophical concepts. But they do not generally fail as proofs because of faulty logic; they fail because they require assent to premises which are insufficiently self-evident, and because their conclusion is of such importance that our predispositions to believe or disbelieve outweigh the argument.

Deduction consists of the derivation of conclusions from fixed premises (axioms) using rules of inference laid down by logic. Nobody who assents to such fixed premises, granted no errors in the deduction, is free to dissent from the conclusions. But we cannot always foresee all the implications of a given set of axioms. For example, many people will innocently accept the following assertion as a major premise:

> All actions which involve taking human life are unconditionally wrong.

They will do so because they have been given no reason to be suspicious. It is a trivial matter to add the following minor premise and deduction:

Defensive wars involve taking human life.
Therefore defensive wars are unconditionally wrong.

Yet very few will accept this conclusion, and since the minor premise is incontrovertible, as is the logic, those who cannot agree will feel forced to qualify their assent to the major premise in order to avoid the unacceptable conclusion.

This example explains why I have not chosen to discuss the various "proofs" of the existence of God in this work: I acknowledge that our prior convictions regarding the existence or non-existence of God are such that even if a valid argument were found with the conclusion "therefore God exists" it would have the result only of forcing the agnostic or atheist to revise his assent to whatever premises had been used, or to quibble about a stage in the logic.

All this may sound rather alarming. Surely, we protest, there are arguments capable of persuading us that we are wrong? The man in the street certainly believes that there are, for he has been raised in a culture exhibiting the highest standards of rigour in both philosophy and science. He is likely, in response to what has just been said, to point to science as the greatest and most powerful counter-example, for there embedded deep within our culture is an entire system of ideas relying upon the fact that we have the means to separate true assertions from false ones. He knows little enough of the mechanics of Newton which predicts the motions of the planets, but he does know that we have been able to perform calculations and build satellites and space-probes which perform with great precision. Being a practical person he can conceive of no more satisfactory confirmation of the truth of science than that it works. It does not matter to him what the theoretical or philosophical basis of his culture is, still less what academics say it ought to be. To put it in the form of a picture, it is as if he, as a leaf on one of the outermost branches of a great tree, were one day to be bent down to the earth and shown the base of the trunk,

RECONSIDERING PROOF

and the beginnings of the roots, which are both his strength and his sustenance, and were to reply "but they are so far from the clean air and the sunshine where I live up above!" Yet if the trunk is rotten, or the roots fail, he will fall with them, and the clean air and the sunshine will not save him.

How would we know if the roots and trunk were rotten, in other words if the fundamental assumptions of our culture were mistaken?

> The most obvious and easy things in mathematics are not those that come logically at the beginning; they are things that, from the point of view of logical deduction, come somewhere in the middle. ... we need two sorts of instruments for the enlargement of our logical powers, one to take us forward to the higher mathematics, the other to take us backward to the logical foundations of the things that we are inclined to take for granted in mathematics.
> Russell, *Introduction to Mathematical Philosophy* I.

The *religious* man in the street is often as impatient with current debates in theology as our former friend is in the nice distinctions of philosophy, because his interest is not in intellectual arguments, but in some kind of feeling for a religious leader such as Jesus. He is content to believe, and insistent on doing no more; he would like to understand the faith, but he does not wish to question it; life probably holds enough questions he cannot answer already. If he is challenged by an unbeliever he will probably reply with a simple restatement of what he believes, may perhaps make a half-hearted attempt to answer typical accusations such as arise from the problem of suffering, but will rest his case upon what seems to be the feeble assertion that although he cannot answer these questions or refute these challenges, he nevertheless remains confident that they can be answered, if only by God.

This book undertakes to begin from what seems to be the *least* hospitable ground of human activity, the rigour and precision of mathematics, and to show how it is not only possible but actually necessary to reverse the formalistic emphasis which places the indigestible axioms which generations of mathematicians have distilled from their study at the forefront of the study of mathematics, and to rediscover the heuristics from which they first arose (what Russell calls the things which come somewhere in the middle from a logical point of view). The same must happen in theology if we are to rediscover a living Christian orientation, in particular that orientation personified in Jesus of Nazareth in his faithfulness to the world, to man, and to God. We too must be faithful to the world in our mathematics, science, and technology; we must be faithful to our neighbours in relationship, community, and world; and we must be faithful to our origin and destiny in purpose, value, and action. Those familiar with the heuristic grounds of mathematics and theology, either by dint of effort or happy chance, too easily commit the sin of retrospective refinement by assuming that others can clothe the skeletons of knowledge we call axioms and dogmas with living flesh. Without that flesh those axioms and dogmas are just arbitrary.

Any axioms or dogmas which we are prepared to affirm without appreciating the concepts which bring them to life are *mere* beliefs, whatever their content. Someone who affirms belief in God without giving any thought to what the term "God" means, or to whether it always means the same thing, holds such a mere belief. We are rightly reluctant to base major claims on mere beliefs because they are inferior to reason, and more prone to error. We recognise that we can deceive ourselves accidentally, and that others may have vested interests in deceiving us deliberately, and we look for confirmation of our claims in less personal kinds of activity. Typically we put our theories to the test, in simple ways by for example trying a new fertiliser on our tomatoes, and in complex ways by building particle accelerators. The availability of tests for our theories (even at vast expense, as is the case with the particle accelerator at Cern) reassures us that we are in touch with reality. Problems arise when the desire for tests starts to become an obsession, so that we withhold our assent from other kinds of claim purely because we can conceive of no experiment which would verify or falsify them.

The logical positivists built such a principle into a system by denying that an unverifiable statement has any meaning. Sometimes it appears as if this principle is aimed exclusively at religious claims, and other values in the humanities, which seem least verifiable. For this reason many writers have attacked the principle of verification without paying sufficient attention to the way it highlights a problem which does need addressing. Once the real nature of enterprises such as mathematics is understood, the question is not whether proposals which cannot be verified or falsified have meaning, but how we are to distinguish *true* undecidable proposals from *false* ones. Why, for example, should we wish to affirm the existence of God while denying the existence of the devil? And on what basis are we to do this?

We live in a culture prone to playing lip-service to the desirability of proof while failing to justify its own assumptions about value, law, purpose, and so forth. We like to think that those assumptions have been justified at some time in the past, and that we can legitimately take them for granted rather as an engineer takes certain formulae for granted when he designs something, or as a mathematician uses theorems proved by others. By thinking this we persuade ourselves that the criteria we apply to new claims are no less rigorous than those that have been applied to our own assumptions hitherto. At its strongest this confidence gives rise to the conviction that everything of importance can be proved according to the principles of mathematical reason, the purest and most incontrovertible kind of proof available. Our fear is that to relax these criteria is to open the flood-gates to a host of irrational claims which will plunge the civilised world back into the Dark Ages.

This concern to avoid *irrational* beliefs is obviously laudable. But the procedures which are supposed to protect us from them have manifestly failed to do so.

Although scepticism has succeeded in putting orthodox religion in the dock, it seems to have had the opposite effect where general superstition is concerned, almost as if by closing the gates on traditional religion by labelling it "irrational" it has forced would-be adherents elsewhere for their comfort and solace, into genuinely irrational pursuits which have escaped the direct fire of the sceptic, and whose practitioners are less concerned for their academic respectability.

Religion, unable to supply experimentally verified evidence for its claims, passes into eclipse before the onslaught of critical philosophy and science. By responding honestly to the demands of the sceptic the religious man traps himself into playing someone else's game by someone else's rules. Less scrupulous religious bodies make no attempt to respond to these demands; they simply reaffirm even more strongly and confidently the same dogmas. And they find that their numbers increase. The paradox, of a highly critical and sceptical world-view producing people eager to surrender their critical faculties to an organisation proclaiming clear and unambiguous truths, may seem stark and unlikely. But it dissolves again once we observe that a "highly critical and sceptical" world-view is based upon and motivated by the need for *certainty*. The offensiveness of undecidable claims is that they may be mistaken, and being mistaken may lead their adherents into error. If, therefore, we are preoccupied with the problem of avoiding being seen to be wrong (and which of us is not?), we can find security either in general scepticism, where in risking nothing we lose nothing, or in the company of others who will supply us with a microcosm of confidence rather like that we enjoy in our wider cultural outlook. The choice between scepticism and sectarianism in a world obsessed with certainty therefore becomes clear. And because life is so uncertain we will adhere most closely to that which offers us either the best hope of certainty, or the most satisfying escape from uncertainty.

SOME PROBLEMS WITH PROOF

Two cases are to be distinguished concerning proofs: in one an attempt is made to prove a hitherto unproven assertion, as when a research student sets out an hypothesis and seeks to demonstrate that it can be inferred from known results; in the other, a student is required to prove a result which has already been proved either as a test of his knowledge (in an examination), or as a means of helping him to understand it better. These two cases might be called *unknown proofs* and *known proofs* respectively, meaning in the latter case proofs known to *someone*, but not necessarily to the person searching for them. Of course, the two cases are not completely distinct, for the student will experience some of the frustrations of the pioneer in his quest; where they do differ is that the student (provided he trusts his teachers, which is not always the case) can be confident that a solution exists, whereas the researcher may be chasing a proof which does not exist (because his hypothesis is flawed), or is not attainable (because the current state of mathematical knowledge is inadequate).

When a known proof is nonetheless unknown to a student he can be set a problem in two ways, either by being asked to prove a formula or result which is given to him in the question, or by being asked to derive some result from a set of assumptions without that result being specified in detail. In the former case he has a clear idea of what it is he is trying to prove, and in the latter he must extract the information that he needs despite the fact that he cannot foresee what that information will be. Consider the following example:

> Tom is four years older than his sister. Two years ago he was twice as old as his sister.

Two kinds of problem can now be posed:

> Show that Tom is ten and his sister six.

and

How old are Tom and his sister now?

Although this problem is so simple, many people will not be able to solve it, especially if it is formulated in the second way. But its simplicity does not prevent us from seeing the difference between the two problems set. As an examinee I will know, in the first case, whether I have finished the problem correctly or not; in the second case there will always be room for doubt and error. (It scarcely needs saying that in simple cases no mathematician will be intimidated in this way, but as problems become more difficult, and their potential solutions more varied, the distinction becomes immensely significant.)

To solve the second kind of problem with known proofs, or discover any kind of unknown proof, I must be able to move about in uncharted territory with considerable confidence. Self-evidently the source of this confidence cannot be a formalised system, since that is what my exploration is intended to discover; it must therefore be my much less tangible mathematical understanding. Yet such difficulties are not confined only to the outer periphery of mathematical research; they occur wherever a problem is encountered which lies just beyond our current intellectual reach, even one such as we have posed here.

The problem is simple enough; everyone understands the individual words, and the concepts in each sentence are straight-forward. Yet in combination they have the power to strike terror into otherwise level-headed people. This is partly because memories of school algebra are often unhappy, and partly because we dislike more than one thing appearing to change at once (because we can really only think about one thing at a time). Mathematics comes to our aid when we introduce symbols for some of the ideas in the question, and write something down which shows how they relate together in a purely *formal* way (by "letting Tom's present age be 'x'"). Once we have done that we cease to be interested in Tom and his sister for a while, and go through a sequence of mechanical procedures which will isolate one of the varying quantities such as Tom's present age ("x") from all the others. In other words, we convert the problem of extracting Tom's age into the problem of isolating a particular algebraic symbol from a given equation.

Memories of school mathematics easily lead us to equate mathematics with searches for proofs and manipulations of numbers. This example shows that it is just as important to be able to convert a problem into a mathematical form (such as an algebraic equation). When most people talk about "mathematics" they mean "arithmetic", and "I can't do maths" usually means "I can't add up". For many, the need for mathematics ends with adding up the grocery bill, and although there is a general awareness that it is also involved in sending rockets to the moon, quite what that involvement is remains unknown. Even moving from arithmetic to simple algebra makes minds "go blank". What is not sufficiently widely appreciated about mathematics, is that we usually have a precise goal *before* we embark upon a solution, and that this goal is crucial in enabling us to formulate the problem properly. In this case the goal is discovering Tom's age, and that concern (to extract information about Tom's age from what we are given) governs our approach to the problem.

This is not a text-book on mathematics, and so I will not pursue the matter further. I ask the reader, however, to take note of several things that may have occurred in reading the last few paragraphs. Unless you are a mathematician you are likely to have sensed a feeling of unease as it became apparent that a *problem* was to be set, followed by a nasty sense of *déja vu* as procedures evoking unpleasant memories of school were mentioned. Any misgivings you had about a book on mathematics and theology were probably made worse. Well, a word of reassurance: that is the first and last such example I shall use, and you need not understand it to appreciate what follows.

We have proved that Tom is ten (in the first form of the

problem), or derived his age (in the second), and since we did not know that before, we have learned something. But there are already, even in this simple example, interesting issues which arise from these answers: first, our proof depends entirely upon the accuracy of the information given to us (*if* the information is accurate, *then* Tom is ten) and makes some rather cavalier assumptions about such notions as "twice" (was Tom *exactly* twice his sister's age two years ago?); second, whatever we have learned must have been buried inside the original information (otherwise we have added something to the problem, which may have distorted it); third, we have used mathematics to convert two rather dull circumstantial pieces of information into two more interesting facts (but where has this new information come from?)

In the following paragraphs I shall highlight as hypotheses popular impressions about the nature of proof which, although they contain germs of the truth, should also raise serious doubts in our minds as to their adequacy. Within this discussion some of the remarks made above will be clarified and extended.

Hypothesis — Proofs generate certain results that are true.

This seems self-evident because we imagine that truths emerge from proofs rather as the perpetrators of murders emerge from detective stories, that is at the last possible moment. We establish or verify the truth of an assertion by proving it, so we think. But a moment's thought will show why this cannot be the whole story. There are countless results which can be derived from a given set of assumptions. Some of them will be interesting, some will lead to further more powerful results, but most, if picked at random, will be worthless, because despite being true (valid inferences) they will lead nowhere. We no more do mathematics by setting off in a random direction looking for results than we undertake a journey without any idea of where we are going.

It is here that the distinction between a derivation of a solution to an equation or problem and a proof needs to be kept in mind. It is certainly true that "Tom's age", for example, may remain unknown until the very last line of the algebra; but it is not true that the direction of the algebra is arbitrary. In fact, it is directed precisely towards that conclusion, and the manipulations which precede it are governed by that end. The problem of producing an algorithm which will reproduce such manipulations is nevertheless very difficult. Although we do not know the exact content of the destination, we know its form. But that makes searching for a mathematical proof much more like an art-form than we usually think likely, for an artist engaged in, say, painting, will be looking for a final product which *does not exist* until it emerges from his labours, and of which he is the sole arbiter of quality and truth.

Sometimes in order to prove something we need to supplement our assumptions. Great controversies surround the augmentation of logical systems with fresh axioms (the Axiom of Choice, for example) because the sheer complexity of the structures under scrutiny prevents us from seeing all the implications of those additional assumptions. They may be innocent and legitimate, but they may also be the equivalent of "All actions which involve taking human life are unconditionally wrong". Nevertheless, if a result cannot be derived without such new assumptions, our desire to integrate the whole system together will produce a strong motive for their inclusion: convictions about true results we wish to incorporate force a modification of the basic assumptions we will allow. (The opposite case, where an unacceptable conclusion forces revision of the assumptions, is not confined to the proofs of divine existence. Most of the pressure upon the place of science in our culture now comes from those who regard the fruits of science such as nuclear weapons, genetic manipulation, and environmental pollution, as equally undesirable conclusions which should force revision of the status of science.)

The firm conviction that an assertion is true gives a mathematician the confidence, the courage, the determination, and the motivation to persist with his quest for a formal proof. Once doubts begin to enter his mind, perhaps because in his searching he encounters serious difficulties he had not foreseen, his mind will begin to turn from the search for a proof to the search for a counter-example (or disproof), and he may easily find neither once the initial drive has evaporated. Amongst other things, this shows that mathematics is an intensely *personal* enterprise in which the commitment of the researcher to his hypothesis plays a large part in his success, quite contrary to the popular image of the mathematician as a dispassionate and detached manipulator of symbols and formulae. The need for such personal involvement arises from the non-mechanical nature of the pathways to proof, and our dependence upon barely-understood processes in the unconscious mind.

A proof is not sought *primarily* to convince us that something is true, but to confirm a prior conviction that it is a true part of a valid system we have already reduced to simple premises, i.e. that it is consistent with those premises.

Hypothesis — A proof consists in the manipulation of formulae.

Our schooling can easily leave us with the impression that what is required in mathematics is facility at manipulating sets of equations in order to generate desired results. But what is it that decides how the formulae are to be manipulated? There certainly are manipulative processes in mathematics, and some of them are difficult and complex, but it cannot be the case that we simply rearrange formulae "willy-nilly", as we might watch the coloured chips and beads in a kaleidoscope. Yet it is equally clear that the formulae themselves cannot prescribe which new patterns are desirable and which not. It is true that mathematics text-books commonly present sequences of manipulations as if they were the essence of proof, but students are notoriously incapable of reproducing those sequences from the countless possible sequences that can be generated. Clearly, there is more to proof than this.

In the first conjecture the emphasis was upon the contrast between conviction and confirmation; here the emphasis is upon the *focus* of the proof, which is different for known and unknown proofs. Where a proof is unknown the researcher is struggling to develop a tenuous thread of connections between his hypothesis and results which are already acknowledged, in which case he will initially be content merely to have arrived at his answer. Where the proof is used in *teaching* rather than in *researching*, this is not the case. Admittedly, schoolchildren *believe* that what matters is the answer, because they believe (with some justification) that the answer will earn them marks. But the focus of education is not upon this or that particular answer; it is upon the processes which can engender understanding. Children complain when, despite having arrived at the correct answer, they receive few marks, because they do not perceive the difference between acquiring a method which will solve any problem, and hitting by happy chance upon a solution to this particular problem. Sadly, teachers and examiners reinforce this emphasis by making it insufficiently clear that what matter are the thought-processes, not the conclusions.

The function of a proof is not just to lead us to an answer, but to enable us to understand better the system of which the answer is only one part. That necessitates a transference of attention and emphasis from learning and manipulating (necessary as they are) to exploring and understanding.

> Hypothesis — Nothing can be deemed to be true unless it has been proved.

This claim can be seen to be false by means of a simple analogy. A school-boy riddle asks "which was the largest island in the world before Australia was discovered?", and the answer is, of course, "Australia". Why should anything different be the case with truth? The fact that we have not beaten a path to its door by finding a formal proof for it does not mean that it is any the less true, although it may mean that our grounds for believing it to be true are rather different, just as it means that there are countless truths so far undreamed-of.

This presupposes a great deal, however, for there is fierce disagreement in philosophy of mathematics about the status of "undiscovered" mathematical theorems. Platonists such as Frege and Gödel were convinced that they were as real and existent as Australia prior to its discovery; intuitionists such as Brouwer and Weyl believed that mathematical objects come into existence when they are discovered. In either case there are problems: if mathematical objects exist independently of us, *where* do they exist, what concept of existence is involved in such an assertion, and how do we come to know about them (since they were clearly not sensed as physical objects); if mathematical objects are creations of the human mind, on what basis are we to choose between rival theories (if we are not to espouse a version of relativism), and how are we to account for the empirical success of those mathematical ideas in physical theories?

To make the matter more concrete, suppose I invent a new problem about Tom and his sister with different numbers, a problem let us suppose which nobody has yet solved (despite the fact that the solution presents no difficulty). Does Tom yet have an age or not? If so, "where" is that explicit age now? If not, where does it come from as we solve the problem?

Gödel showed in 1931 that there are true statements in any sufficiently rich formal system which can be proved to be unprovable (contradictory as that may sound). A great deal has been written on the implications of this result, and I do not propose to add to it here other than by making two observations: first, that although the Gödel results are of great interest the Gödel sentences themselves are not; second, that however extravagantly they are interpreted, their most important implication is that *truth is not contingent upon proof.* Proof does not create truth, and is therefore truth's servant, not its master.

Hypothesis — Proof demonstrates that truth is based upon more elementary truths.

Some of the problems with this hypothesis follow from the previous discussion. But more can be said. The impression proofs create is that from extremely simple statements, which are so self-evident that they require no proof themselves, other more complex statements can be derived which, because built out of these elementary truths, must therefore themselves be true. This is, in part, what proof does. But in order to satisfy the strict criteria that such inference entails, in order so to speak to make sure that there is not the tiniest chink of looseness through which inaccuracy could creep, proof must therefore be content with *adding nothing* to those truths which are contained in the premises upon which it operates. Proof, in other words, *preserves truth*, but does not create or extend it.

We have already touched on these points when discussing the "Tom's age" problem, but this is an extremely difficult concept to assimilate fully, for it is obviously our everyday impression that the theorems or assertions we prove are *more informative* than the assumptions from which we prove them. The question is, where has this additional information come from? It cannot come from proof itself, for in that case the proof would have added something to the assumptions it works upon (something *unacknowledged* and hidden), and therefore potentially have introduced sources of error into the argument. In that case it would not simply have preserved truth, but augmented it (perhaps with falsehood). In fact the rules governing mathematical proof are designed to exclude the merest whiff of new material. If we accept that they are successful (as I am sure they are), the problem remains of where this apparently new information comes from.

One answer is simply to say that a proof merely makes *explicit* what is already and always *implicit* in our assumptions. Proof so understood makes truth *accessible* to us without creating it or distorting our assumptions. That seems extremely satisfactory until we realise that in that case *proof only ever tells us what we have already assumed*. And that seems catastrophic for the entire rational enterprise, for it implies that the only things we can prove are the things we tacitly assume in our premises (yet the whole purpose of proof is ostensibly to show how *new* truths can be linked with our extremely *modest* assumptions). Now it appears that far from being modest our assumptions must actually *contain implicitly everything we could ever wish to prove from them*. Put at its bluntest, we must assume everything we want to prove. But of what use is a proof which only shows us what we (implicitly) already know? In particular, attempts to prove the existence of God reduce to a tautology, for the only way we could ever prove God's existence would be by assuming it in our premises (thus defeating the object of the exercise). (I refer the reader interested in this problem to Michael Dummett's brilliant essay "The Justification of Deduction", in *Truth and Other Enigmas*.)

Proof does not show that truth is based upon more elementary truths; it shows which truths are implicit in the elementary truths which we assume. And those implicit truths are built into the more elementary truths by the processes of axiomatisation. The purposes of axiomatics therefore include making explicit the truths which might otherwise remain undetected in the larger system for which a particular set of axioms is a basis.

> Hypothesis — Proof begins from elementary propositions which are self-evident.

The elementary propositions of mathematics are called *axioms*, but the nature of an axiom is obscured by the common usage of the word "axiomatic" to mean "self-evident" or "fundamental" because, as the Russell quote illustrates, axioms are *not* fundamental self-evident truths. This becomes clearer if we consider a little history to elaborate some of the remarks made in the chapter on doubt.

It is well known that physicists during the late nineteenth and early twentieth centuries made fundamental discoveries about the constituents of matter. Following the Greek term for indivisible minimum, they called these particles *atoms*. It is less well known that there was a school of philosophy called atomism dedicated to the discovery of the indivisible minima of logic. Bertrand Russell was an early adherent of this school. Just as physicists quickly realised that their so-called atoms were not indivisible, so philosophers came to doubt the status of atomic propositions. Nevertheless, the idea stuck, and the usage of "axiomatic", on which we have remarked, is really an unfortunate equivalent to "atomic". In general we now expect to find ultimate constituents neither of matter nor of language and logic: atomism has been abandoned. But the impression that mathematics is based upon elementary propositions persists. In fact, axioms are sets of propositions *sufficient to prove all the theorems in a system*. They are not unique sets, for it has been shown that there are many differently composed sets of axioms for any given system; nor are they immutable or self-evident, for we can in fact *choose* which axiomatisation we will use.

The generation of a mathematical system therefore follows the following pattern: first, a set of interesting similarities between apparently different phenomena is noticed; second, some of the general properties of these similarities are formalised using mathematical symbols and equations; third, some of the connections between the properties are charted, and theorems are put forward and proved using other, apparently more self-evident results; fourth, over a long period of time (perhaps centuries), the fundamental assumptions implicit in the properties under study crystallise out, and it becomes possible to put forward small sets of such properties from which (it is hoped) all other results can be derived. Once this axiomatisation has been performed the hope is that the *economy* it affords will permit a more efficient cross-referencing of the system from which new, hitherto unnoticed results may be put forward and proved. The entire process (abstraction, formalisation, axiomatisation, deduction) is governed by the mathematical community's understanding of the structure under scrutiny, and is motivated by their collective perception and intuition of the *fruitfulness* of the ideas it contains. The new system acts as a gateway to as-yet-unforeseen possibilities.

> In mathematics, the greatest degree of self-evidence is usually not to be found quite at the beginning, but at some later point; hence the early deductions, until they reach this point, give reasons rather for believing the premisses because true consequences follow from them, than for believing the consequences because they follow from the premisses.
>
> Russell and Whitehead, *Principia Mathematica*, Preface.

Therefore proof, far from being a means of showing that mathematical truth is based upon elementary propositions, is really an attempt to reconstruct the overall perception of the system of ideas which gave rise to the axioms. Instead of being orientated downwards into minute particles of knowledge, or backwards to reproduce results already proved in the past, proof aims to draw together the power which the axiomatic method confers and the vision which mathematical ideas afford so that armed with these tools new ideas leading to yet more visions may be discovered without our inadvertently introducing contradictions by being insufficiently aware of their elementary connections.

These positive and negative observations about the nature of proof contrast sharply with the popular *impression* of proof as a means of progressing from simple and obvious certainties and truths to more complex, but apparently less obvious statements. Particularly disturbing is the suggestion that anything a proof demonstrates is already contained in a hidden or implicit form in the assumptions from which it begins. That seems to convict proof of deception on a massive scale, for it implies that however persuasive a proof may be — however far it seems to take us from simple assumptions to profound conclusions — all it really does is to show us some aspect of our assumptions which we had not noticed. Just these conclusions led Wittgenstein to the famous observation, made in the *Tractatus*, that all the propositions of logic say the same thing, to wit nothing. All logic is a tautology, for once proofs add anything to their assumptions they run the risk of adding errors, and the reliability of logic resides precisely in our confidence that it does not do this: in a sense it is impossible to make mistakes in logic.

THE VALUE OF PROOF

The distinction between formal and non-formal worlds gives us a clue to the central functions of proofs. Suppose someone experiences the problem of finding it difficult to express what he *means*. Somehow the words just don't seem to do justice to his understanding. He is at a loss to say exactly how they fall short, but he is acutely aware that they do. The dual control language and understanding exert upon one another makes this absence of appropriate words extremely frustrating. The desire to give expression to what has been understood is partly a desire to formalise it for oneself (to know what it is one knows, so to speak), and partly a desire to communicate with others. But there are further functions which such expression serves, and which proofs share. For the sake of a convenient scheme they can be called extension, regulation, preservation, formalisation, communication and clarification.

1 — Extension. Proofs extend our concepts by means of tacit integrations. If I work through a text-book proof mechanically I learn nothing. But if I ask myself the significance of each step and how it relates to broader mathematical concepts; if I approach the same result using diverse proofs which employ different techniques and concepts; then I will make my understanding of the result richer and by so doing enlarge the concept it represents. The distinction between the formal proof (consisting of symbols and logical operations) and its non-formal meaning must be borne in mind: proofs do not interpret themselves; we must integrate them to their focal meaning.

2 — *Regulation*. New hypotheses often just "occur" to us as our unconscious throws up some undreamed-of idea. The question arises not only of how important the result is, but of whether it is consistent with and derivable from existing results. In other words, is it part of the same conceptual system; are we still speaking the same language when we talk about it? Here proof (when it is available and found) enables us to show that the new result can be derived from other accepted results, and as such forms part of the same language. There is both a formal coherence and a conceptual coherence.

3 — *Preservation*. To complement its regulatory function, proof also serves to ensure that such truth as is inherent in the premises of a system also resides in its theorems. Of course, the question of what that truth is, and of the notion of truth appropriate to mathematics, remains problematic. But we can assert with confidence that whatever the answers to those questions are, proof preserves that notion of truth.

4 — *Formalisation*. It may be asked why we bother to give expression to our understanding at all if understanding is so much richer and essentially ineffable. There are a great many answers to this (some of which relate to the other facets of proof):

> (a) I can keep my entire train of thought before me at once, whereas I can only "think" one thought at a time (i.e. to think one thought I must forget or store all others), and powers of recall are not what they might be;
> (b) I can provide a framework to cross-reference my ideas for coherence and continuity;
> (c) a particular form of words or symbols may encapsulate an idea particularly appositely, as in a quotation, and perform the function we shall call mechanised intuition by enabling me to call up meanings more efficiently;
> (d) whereas in my mind I may delude myself that my thoughts are coherent, being able only to think one thing at once, on paper the

contradictions in my thoughts become more obvious; they force me to be precise, technically and aesthetically;
(e) formalisation fixes our thoughts outside time, and so enables us to check back over our own work in the future and to obtain access to the work of others;
(f) the sheer act of expressing oneself on paper forces ideas to become sharper and more coherent, partly because it makes us aware of the deficiencies in our arguments and gaps in our knowledge; I may think that I know something, but when I come to try to write it down I find that I do not;
(g) by presenting us with sequences of ideas which can run counter to our train of thought of the moment, old writings of our own or the writings of others may provide the material for new integrative leaps or insights, since otherwise we would be entirely dependent upon the ideas our own unconscious chose to juxtapose;
(h) by showing us the path we have traversed our proofs can lead us to short-cuts as we realise that there was a far easier way from statement one to statement thirty involving far fewer steps; but often such short-cuts only become clear after we have laboured to traverse the longer path;
(i) because we know more than we can tell and tell more than we can ever know, our formalised ideas can be sources of new discoveries for other people who, able to integrate them into their own experience and understanding, can achieve insights which would not otherwise occur since the author of the words or proofs lacked access to the material;
(j) awareness of successful and unsuccessful lines of enquiry from our own and others' past can make us more efficient.

5 — *Communication.* Many of the features of formalisation are necessary for communication. But communication is not the same thing as formalisation, since manifestly I can say and write things which are not understood. Proofs therefore can act as a challenge to others to try to understand something which presently they do not understand. One can say "yes, I see that *formally* this is a proof, but I still don't really *understand* it". The force of communication stems from the sense that someone here has something to say, that is that someone is trying to say something *to me.* That conviction alone can lead me to struggle to understand. For example, when I first read Polanyi's *Personal Knowledge* as a mathematics under-

graduate (on a personal recommendation of someone whose views I respected) I found it immensely hard going. Nevertheless, Polanyi did communicate in very clear ways the fact that *here was something important*; he did what he had attempted to do, he generated a heuristic field (a field of finding-out) which drove me to greater efforts in order to understand it. It is an illusion to suppose that valuable things are immediately accessible (or that we would always recognise value when we encountered it).

6 — *Clarification.* We all make the mistake from time to time of thinking that we know something (such as the route to a house), only to find that we do not (having driven round for hours looking for it: "I'm sure it was just here!" we say). Sometimes in speaking or writing something we find ourselves confronted with something quite new: "Good heavens! Fancy that. I never knew that before!" Equally commonly we find that what we thought we knew has suddenly become unclear. In teaching my wife to drive I suddenly found that I couldn't tell her whether to engage handbrake before gear or vice versa. How absurd! I do it every day. Or suddenly one looks at a word one has written: surely it isn't spelled like that? It looks so odd! And nothing short of looking it up in a dictionary will resolve the matter (but usually we write it without thinking twice). We clarify our ideas in mathematics by seeing whether we can provide proofs for conjectures, or solve associated problems, since the ability to provide a proof or a solution is a good indication of whether we understand.

Proof is therefore a means of maintaining the coherence and consistency of mathematical knowledge by demonstrating that only certain combinations of statements are compatible. What proof cannot do is to demonstrate *which* sets of compatible assertions are both true in relation to one another, and true of the world. In Wittgenstein's powerful phrase, we cannot say in language how language relates to reality.

The way out of the obvious resulting dilemmas (how

then do we know that anything we say is true, or which of the things we can say are true?) has been sought by all philosophers. Some, such as the atomists, have sought a solution in irreducible and undoubtable elementary propositions, but their attempts have failed because even these self-evident truths can be shown to be arbitrary, and because we now have a multiplicity of different *logics*, each built up from different, mutually incompatible sets of elementary propositions (axioms), which must themselves be decided between.

Significantly, philosophy of mathematics grew up alongside philosophy of language. Frege, the great German logician of the turn of the century, Bertrand Russell and Wittgenstein all made major contributions to both. In both fields the insolubility of the problem of relating mathematics and language to reality has figured large, and has always threatened to result in one of two unsatisfactory conclusions. In the first, we feel honour-bound to condemn ourselves to silence on the subject: "nothing will do as well as something about which nothing can be said" (*Philosophical Investigations* 304). In other words, since we cannot say how language relates to reality using language we should abandon the attempt. Clearly language does relate to reality, but we can say nothing about how it does so. In the second, we acknowledge that language relates to reality in an arbitrary manner, and conclude that its meanings are mere conventions which we could in theory revise, but in practice do not.

Every attempt to resolve these dilemmas in impersonal terms by reference to impersonal objective criteria will and must fail. There is no impersonal Archimedean Point to act as a criterion of demarcation between the inventions of the human mind and genuine insights into reality. Thus, far from embracing Kant's transcendentalism as a cure for this apparent problem ("transcendentalism" meaning a philosophy which searches for and claims to some extent to have found a fixed point of rest from which to view things in the perfect light of eternal truth), we should instead embrace its inescapable partiality and set about investigating in a determined way what implications that has for our knowing and being. I shall argue, in other words, as one pole of my thesis, that *absolutism* is an enemy of truth and not its ally, for absolutism blinds us to the limitations of our vision and the culture-ladenness of our knowing, and impairs our progress and growth as a result. However, this is only one pole of my thesis, and it must be understood in conjunction with its counterpart, which denies that *relativism* is the logical outcome of such a program, and puts forward a third way which reinstates the genuineness of our knowledge of the truth without reinstating absolutism into the bargain.

POST-CRITICAL PHILOSOPHY

Davis and Hersch, in their book *The Mathematical Experience*, present their readers with an Ideal Mathematician who, as it happens, is researching into non-Riemannian hypersquares. He is engaged in a dialogue with an Information Officer attempting to discover what precisely his research is about. It proves impossible for the Ideal Mathematician to communicate with this man (not surprisingly), but the extreme example raises the following important question in the present context. What convinces the Ideal Mathematician of the existence of non-Riemannian hypersquares? It cannot be sensory experience, experimental evidence, or human testimony. It must be his personal experience of their conceptual unity and intrinsic intelligibility, an experience which he shares with the handful of others similarly engaged in this esoteric field. In other words, he affirms their existence because he can conceive of them, not just as an isolated individual (that would be a recipe for subjective delusion), but in conjunction with a community of enquirers whose corporate conception gives rise to a consistent and intelligible formalisation, a language in which the mathematics of non-Riemannian hypersquares can be discussed. That is emphatically not to say that this small community *makes* these entities exist by its labours, or that such entities would not exist as objects of possible knowledge and study without them, but that their existence would be closed off from us all; we would neither have reason to believe in their existence, nor the means to speak of their attributes. The Church stands in an exactly similar position in relation to the existence of God.

That does not resolve the question of existence posed earlier, however, for we can perfectly easily enter upon a long and detailed conversation about blue unicorns and publish learned papers on the subject without in any way thereby necessitating the existence of such creatures. In other words, a community of enquiry dedicated to articulation of an adequate language is only a necessary and not a sufficient condition for the objects it describes to be known to us. But more than this can be said. Our Ideal Mathematician friend will, in practice, scarcely ever give a thought to the question of the *existence* of non-Riemannian hypersquares other than when challenged to do so by someone such as the Information Officer. He is totally immersed in his study, and it would be absurd for him to imagine that he was immersed in study of non-existent entities.

This is just the point that John Calvin had in mind when he insisted upon reversing the mediaeval sequence of questions *quid sit, an sit, quale sit,* as T. F. Torrance has so often pointed out. It is also the point that many modern philosophers have made in observing that to ask about the existence of something, say "p", in isolation, is meaningless unless the *sign* has some reference or some sense. Like the Ideal Mathematician, we must start from some more-or-less definite concept "God" before the question whether he exists can have any meaning. Therefore *description* is necessary for the nature of the object to be intelligible. Thereafter its intrinsic nature will be able to convince us whether or not it also exists other than as a creation of the human mind. Of course, this involves making a personal decision, but this is just what I do with claims about blue unicorns, ghosts, U.F.O.s, astrological predictions, and things that go bump in the night.

The churches have unfortunately become so concerned with the question *whether* God exists that they have neglected exposition of his nature. Consequently the God whose existence they are concerned to defend slowly dissolves in their corporate consciousness, and they find themselves arguing for the existence of someone whose nature they are scarcely able to articulate. But the question of the nature of God can only be answered within a highly sophisticated living formal system served by dedicated men and women who, like the Ideal Mathematician, scarcely question the existence of the object of their study, so real and alive is it to them. In repeating mathematical proofs we acquire fluency in mathematical discourse and obtain access to mathematical realities. In rehearsing theological arguments we acquire fluency in theological discourse and obtain access to theological reality.

It is our ability both to create and to recreate in our minds conceptual fields which reveal to us the nature of intangible objects which makes us unique as human beings, able to apprehend God and the deep rationality of the universe. Yet only by *first* becoming intimately familiar with such concepts are we able to resolve the question which are *mere* creations of the human mind, and which recreations of realities existing apart from us. Acquisition of the skills necessary to make such decisions occupies us for a lifetime, but is especially evident in children who are constantly working out which aspects of the worlds we place before them are real and which imaginary. Many adults of course are unable to face reality, and prefer to live in phantasy worlds which offer illusory and temporary comfort.

Michael Polanyi articulated the personal criteria by which we are enabled to make these decisions in what he called *post-critical philosophy*. There are four problems which a post-critical philosophy must resolve if it is to fend off the critical challenge once we admit that conflicting systems of thought which cannot be connected logically cannot be distinguished and refuted logically.

The problem of truth — How are we to distinguish true systems from false ones?

The problem of circularity — How can the charge that allegiance to any such system is circular and subjective be refuted?

The problem of conversation — How can conversation between mutually incompatible systems be sustained?

The problem of conversion — How can conversion from one system to another be accounted for and achieved?

The problem of truth is the most difficult of these, and our response must be delayed until we have developed some further ideas, but the other three issues can be tackled immediately.

CIRCULARITY

Two absolutist objections to post-critical philosophy are that the systems it envisages are *circular*, and that allegiance to them is *subjective*. Polanyi admits the first charge, but strenuously denies the second. "Any enquiry into our ultimate beliefs can be consistent only if it presupposes its own conclusions. It must be intentionally circular." (*Personal Knowledge*, p. 299.) The objector believes that there are systems which are *non*-circular, that is which start from incontrovertible premises and move logically to incontrovertible conclusions. He also believes (by definition) that such linear systems are *objectively true* and completely *impersonal*. His allegiance to them involves him in no decision-making; anyone similarly placed would automatically accept them as true. As objective, impersonal truths they must therefore be *universally* true, and as a necessary extension all claims not logically compatible with them must be *false*. The reason why some people hold to these false claims can and must be accounted for in terms of deliberate deception, malice, ignorance, stupidity, conditioning, or some other pejoratively interpreted cause.

What is most striking about these certainties is the security that they offer their adherents, the conviction that they are in touch with the truth and nobody else is, in effect that they are right and everyone else wrong. Such sureness easily becomes self-righteousness, an inability to see or imagine that you might be mistaken; it can become a recipe for regarding yourself as knowing how to distinguish truth and falsehood, good and evil, even the saved and the damned, purely by applying impersonal rules.

As we have seen already in outline, there is a natural opposite to this steadfast certainty about the truth, namely an equally strong conviction that *nobody* knows the truth, even that there is no truth to be known. The certainty of knowing the truth gives way to the certainty of knowing that there is no truth. In such a view everybody is right and nobody is right; we must live and let live, because there is no basis upon which anyone can decide which way we should live or which ways we should not — as long (we add hastily) as we do not interfere with anyone else's freedom to do the same.

Critical philosophy, based upon the principle of methodical doubt, therefore gives rise to two equally illegitimate children. One, *absolutism*, assumes (irrationally) a position which rescues it from the shapelessness and emptiness of life made devoid of meaning by our inability to prove anything according to the canons of scepticism, and defines its own version of rationality within its chosen structure while denying that a choice is involved. The other, *relativism*, assumes (irrationally) a position which removes from us the need to take any risks by affirming a strong version of the view that truth either does not exist (except as a figment of our arbitrary language), or that even if it does exist we cannot know what it is (except by an arbitrary act of will).

Notice how *stable* these positions are, especially when under attack. The absolutist deflects all assaults by reference to his canonical rules (as in biblical fundamentalism); the relativist replies to all aggressors with the demand "prove it!" *Neither the canonical rules nor the desirability of proof can be questioned.* They are both in this respect internally consistent. That is, they both remain true to the *axioms* of their own system. But they do not acknowledge that those axioms (like all axioms) are *chosen* rather then *found*. That is, they each claim that the axioms which form the basis of their systems are *self-evident truths* which no "reasonable" person could deny, where "reasonable" is defined to be equivalent to "agreeing with these axioms". In other words, a reasonable person is someone who believes in and agrees with the axioms of system X, and system X will therefore be regarded as true by everybody who is "reasonable". How could it be otherwise, from the point of view of an adherent of the system?

What is to be objected to about these stances is not that they are circular — indeed, we have made the point that all systems are circular, even in mathematics — but that they *pretend not to be*, and make their principal charge against other systems that *they* are circular. They therefore practise the *transcendental pretence* of being based upon firm and unshakeable universal foundations. The cardinal principle which stands against this position is as follows:

> Cardinal Principle 1 — There are no absolute impersonal foundations.

At first I intended to call this book *Being Without Foundations*, with the intentional *double entendre* that we *are* without foundations, and therefore being *without foundations* we must learn to *be* without foundations. Unfortunately such a title would probably give quite the wrong impression, but the concepts involved are nevertheless appropriate for the whole enterprise.

Mathematics is an educative and developmental science whereby we enable successive generations to become acquainted with powerful systems of concepts in order that they should be equipped for the exploration and development of new ones. The most essential ingredient of mathematics is therefore *vision*. With vision comes orientation towards that which is important and true, and with that orientation comes the development of the mathematician which brings with it fruitful new ideas.

The crucial word in what has just been said is really *towards*. One can be orientated *towards* something without having it firmly in view, or pretending to have comprehended it. One can be orientated towards the truth without claiming to have grasped it and encapsulated it in a formula. What matters is that one is on the right road and heading in the right direction; and what guides you in that journey is your vision of the goal which you seek and the scope and power of the system of ideas you espouse. The distinction between transcendental philosophy incorporating the transcendental pretence, and a transcendent vision, is that in the former we seek and believe ourselves partly to have found a final fixed and absolute position from which to assess all theories and claims to truth (as in scientism), whereas a transcendent vision involves no such pretentious claims, merely a willing reliance upon a beacon which guides us towards the truth however insecure the ground or path which we tread.

The visions we have and the orientations we take up are inevitably governed and limited by the systems we choose in order to make sense of the world and guide us forward through life. The question therefore becomes: if logical criteria are not adequate, what criteria are there which we can use in making so vital a decision?

One of the many answers to this question is that we should use *religious* criteria, that is that all the criteria offered us by political, social and economic arguments are inferior, as are those orientated towards pleasure, profit, the quiet life, and so forth. Of the many sets of religious criteria, one consists of pointing to the person of Jesus of Nazareth, proclaiming that he is the Christ, the anointed and chosen one of God, and answering the question by saying that *his* criteria are the ones to be followed in making this decision. That involves saying that the vision which he had of God, Man and the world, and the orientation of his mind towards God, Man and the world, are the *ideal* vision and orientation, which we should each take up. It involves saying that Jesus is "the Way, the Truth, and the Life", and responding accordingly.

It will be objected that such an allegiance, if it is wholehearted (and it is difficult to see how it could be anything less), would involve a huge *risk*, and *entering into a system which is circular*, for a person such as the system claims Jesus to be will prove to be *personally self-authenticating*, i.e. the centre and basis of the system will be the person of Jesus himself. But this is entirely consistent with all that has been said already: all systems are circular, and therefore all systems involve risk, even those which pretend not to be circular and therefore not to involve risk. From inside the "Jesus system", however, it seems that since the whole of life involves risk it must be the case that God designed life to involve risk. Therefore, since God is good, it must be *necessary* for us to take risks if we are to come to fullness of life. And the biggest risk of all is the risk God forces upon us, that we will choose the wrong system to base our lives on. That raises the question of the nature of tragedy, which will concern us in the final chapter.

The objections of the relativist are different. If all systems are circular, then since there is more than one system, must we not in honesty adopt some version of relativism, since any system is potentially wrong, in the absence of satisfactory evidence of its correctness? Put much more succinctly this could be phrased as, "those who recognise that they do not know the whole truth should conclude that they therefore may know none of the truth". Or, in yet another form, the objection is that if all systems are circular we must abandon hope of knowing the truth. This is another form of the Fallacy of Democracy.

A clue to the solution of this problem comes from Polanyi's distinction between subsidiary and focal awareness, and the consequent distinctions we can draw between our formal and non-formal worlds. Polanyi perceived that understanding and truth are ineffable, that neither understanding nor truth can be encapsulated in words. In this he echoed many themes in contemporary philosophy, notably Wittgenstein's realisation that we tell whether someone understands a concept by the way he uses a word associated with it, and Frege's insistence that a word is not in itself either a meaning or the name of a simple mental image. Yet we *need* words and other formal objects and actions in order to communicate. Indeed, we assess the extent of someone's understanding by the range and power of their usage of formal expressions in novel and unexpected ways. Concretely, in mathematics we set problems which can only be solved in part by rote learning; there is always one part of a question which stretches and twists book-work, and which can be solved only if the concepts involved have been mastered sufficiently to provide overall control of the manipulation of the formal terms.

CONVERSATION

Polanyi argued that the formal expression is something we must *dwell in* in such a way that it becomes *transparent* and yields insights into what it conveys by allowing us to *focus* upon that meaning. When I dwell in your words and actions I am able (to a greater or lesser extent) to gain access to what you *mean* and to the *understanding* which shapes what you mean. In other words, I gain access to your orientation towards the world, your mind.

The inability of the formal to encapsulate the non-formal ensures that *we know or understand more than we can tell*, and the extraordinary diversity and richness of the minds which encounter our words and actions ensures that *our words tell more than we can ever know or understand*. This partly explains such experiences as reading some of our own words and finding that they lead us to new insights which were far from our conscious minds when we wrote them. For example, "I didn't know that I thought that until I heard myself saying it".

The power of discernment, of seeing the meaning of the formal by dwelling in it in such a way that it becomes transparent and discloses the non-formal world from which it came, can yield us knowledge of the truth which we cannot put into words. Such knowledge or experience may never take a form which can be expressed in statements, but it can shape our orientation to the world, and as such govern the kinds of people that we are. In its highest form of expression, in the being of God, such truth cannot be expressed adequately in words at all, for words are the creations of men. But whereas the truth of God cannot be expressed in human words, the Word of God can be expressed in human life, for human life is not the creation of men, but of God: "not of the will of the flesh, nor of the will of man, but of God; and the Word became flesh, and dwelt among us, and we beheld his glory, glory as of the only-begotten Son of God, full of Grace and Truth".

Just as we need to understand a language and the culture from which it springs in order to use its formal expressions properly, and just as we need to understand the systems of ideas and abstract structures of mathematics if we are to wield its formal symbols effectively, so we need to understand the nature and purpose of life if we are to live it fully. As in the first two cases, where lack of understanding inhibits expression and leads to mistakes and incoherences, so with life the absence of a *supervening vision* (as Roger Sperry has called it) leads to fragmentation and dissipation. That will involve turning from an orientation downwards and backwards in search of roots, to focus our minds forwards towards the purpose of humanity, and upwards towards the being of God. It involves being without foundations.

This program — to be without foundations — will constantly meet the question "why do you believe or act or conduct your life as you do?" The question throws the protagonists into the middle of the following dilemma:

axioms arise from systems, not systems from axioms; therefore to understand the arguments which follow from statements taken within a system to be axiomatic requires that those axioms are understood; but the axioms only become comprehensible from within the system in which they arise; from outside they seem unacceptable, even **false;** therefore, since apparently the only way a system can be understood is by working one's way through its axioms, proofs and theorems (as in mathematics), it seems as if nobody can ever understand someone else's system of ideas. Put summarily the paradox is as follows: to understand anything (I say) you must understand everything (I say); but you can only understand everything (I say) by first understanding something (I say). This seems impossible. Yet this dilemma is central to education and nurture: how is understanding to be acquired and taught?

CONVERSION

A central problem in education is how to impart appreciation and understanding of a Gestalt. How is it possible to share a vision of the whole with pupils who lack both the experience and the knowledge to see it for themselves? The problem is that possession of such a focal vision is necessary both to *motivate* further exploration, and to *integrate* otherwise apparently unconnected or worthless subsidiary elements in our study. It is as if unable to grasp the whole we cannot order the parts, and unable to see the point of the parts we cannot grasp the whole.

For small children the problem is overcome in play, exploration, and a beautiful indifference to being wrong. Older children and adults need to rediscover each of these features of learning if they are to have the confidence to permit new ideas to impress them with their richness, and to come to know them intimately. Just as a scientist proceeds with his work until he encounters a problem which forces him to examine again some of his fundamental assumptions before he can go further; just as children practise skills repetitively by inventing games which test and stretch them, and so force back the boundaries of their ability; so every learner must have the confidence to approach his subject as if nobody had ever been that way before. He must not be afraid to make the same mistakes as his forebears, even if he keeps reinventing the wheel. What matters is that he has acquired a skill. Far too many pupils imagine that they will pass their exams because their teacher can answer the questions. Teachers and pupils should constantly bear in mind that even epoch-making discoveries only emerge after strings of errors have been eliminated and hosts of blind alleys explored. Even great discoverers are reluctant to publish their mistakes (which does not mean that they did not make any).

There is always a danger, reinforced by the desire among teachers for it to be true, that pupils will believe the phantasy of "right first time", that everyone but them solves problems or understands theories immediately. They then start to see mistakes as signs of deficiencies only in themselves, failings not shared by peers, great discoverers, or, especially, teachers. Consequently, "being wrong" is seen and felt as a sign of personal failure. Contrast, for example, Seymour Papert's experiences with the computer language LOGO (described in *Mindstorms*), in which the educational philosophy relies upon the child discovering that he or she is wrong by the simple fact that the program fails to work. No admonition is issued; the child is simply encouraged to try again. We need, following Papert's example, to rediscover the element of play which makes ideas appealing, not intimidating enemies but enjoyable friends.

The learning of our early years is not inhibited by fears that we may fail, be punished, or lose face if we make mistakes. Children are allowed to explore language inexpertly as they search for adequate expression, and research has shown that parents instinctively offer correction only where verbs are wrongly used, preferring to *show* how language should be spoken, by example, rather than by specific formal teaching. None of us learnt our mother tongue from a grammar, but most of us are required to learn a second language from a grammar. Why is this? Is it necessary? If we can "pick up" our native language by practice, why do we not learn second and third languages the same way? I vividly remember being taught French by a woman we were all afraid of. Lessons were associated almost exclusively with being wrong, and being punished for being wrong. That is no atmosphere in which to teach anything, for fear is the greatest inhibitor of learning. The paths to fluency and self-confidence will be blocked by destructive anxiety and self-consciousness. Ask yourself how many English irregular verbs there are. Do you know? Had the question ever occurred to you, other than as an exercise in an English lesson? Of course not, for we do not *need* to know which English verbs are irregular; their conjugation is second-nature to us.

> Cardinal Principle 2 — We have only understood a language and its grammar when we can dispense with the grammar.

The only children who can understand English grammar are those who do not need it: when the language has been mastered the grammar is easy; when it has not been mastered it is useless. The same is true of mathematics. Children who understand algebra and geometry can manipulate the symbols effortlessly because they have access to the supervening vision which guides and governs that manipulation; children who do not understand geometry and algebra do not know how to perform the appropriate manipulations, and so fail to acquire understanding by practising them. They simply become more and more disheartened as they get bogged-down in pages of meaningless symbols.

> Cardinal Principle 3 — If we do not know where we are going we will not arrive.

That is not to say that proof has no part to play in teaching mathematics, but that purely *formal* proof is unlikely to communicate understanding of anything to anyone. The kind of proof that is helpful is the kind of proof that arises from and accompanies *play*. By "playing around" with mathematical objects and ideas we enrich the conceptual world surrounding the formal terms and gradually acquire understanding. An A-level chemistry master taught me that the most valuable way to work, granted that facts do have to be learned and text-books to be read, is to sit down with a blank sheet of paper and see what you can deduce from what you know. Usually one discovers that one knows more and less than one thought. Once something has been fully assimilated and understood the "foundations" become unnecessary.

> Cardinal Principle 4 — We have only understood a theorem and its proof when we can dispense with the proof.

"Understanding" here, and in Cardinal Principle 2, should not be equated with "knowledge". It is perfectly possible to *know* a theorem and use it appropriately without knowing its proof, and to know a theorem without understanding it. *Understanding* a theorem involves knowing more than we can tell about it; it involves having acquired something of the *conceptual penumbra* that gives the theorem its richness. Such acquisition of understanding, such learning, permits us to use words and theorems, and to appreciate the use of words and theorems, in immediate and intuitive ways which do not depend upon the performance of some preparatory analysis (as, for example, when I have to think where to place the verb in a German sentence). Words and theorems internalised to this extent provide us with the means to communicate with one another using what might be called *mechanised intuition*. Hearing a word such as "chair" or "circle" in some context *immediately* conveys something to us, and what it conveys to us is far more than words alone can express.

> Cardinal Principle 5 — Meaning is not reducible to language.

The expression "mechanised intuition" demands some explanation, because it seems self-contradictory. Consider the fact, frequently commented on in philosophy of language, that although I cannot give you a precise account of what my understanding of the word "chair" consists in, I remain completely confident that I can recognise a chair when I encounter one, and can distinguish appropriate from inappropriate uses of the term in spoken and written English. The word "chair" effectively *calls up* my intuition of "chairness" without my engaging in any process setting that "calling up" in motion. Words effectively "switch on" our intuitions of their meaning in ways beyond our control.

> Comprehensive knowledge of the reference would require us to be able to say immediately whether any given sense belonged to it. To such knowledge we never attain.
> Frege, *On Sense and Reference* in Geach, p. 58.

If we knew enough about the structure whose axioms we study we would not need them; it would be immediately clear which assertions were theorems in that structure and which were not. We would not need to have recourse to proof. In ordinary language we approach this ideal much more closely than in precisely formalised language (as the quotation from Heisenberg given earlier intimates).

Only by resorting to some fairly extreme methods (such as repeating a word over and over again until it becomes meaningless, temporary as even that proves to be) can I prevent myself from understanding a word which is familiar. Therefore my understanding of the word "chair" is mechanical; but it is not *only* or *merely* mechanical because I can subject myself to influences which will enlarge or diminish my range of concepts, and because decisions about which influences to subject myself to are not reducible merely to mechanised intuition. By subjecting myself to mathematical influences I intend to learn more mathematics (i.e. acquire more mechanised intuitions about mathematical ideas); with religious influences I intend to learn more about life and God. Moreover, my concepts are changed whenever I encounter new sentences offering me new connotations of the words I use.

Provided that a symbol is connected to the same concept in the minds of A and B then by uttering [speaking, writing, drawing, projecting on a screen etc.] this symbol, A can evoke the concept from B's memory into his consciousness — can cause him to 'think of' this concept in the present.
   Skemp, *The Psychology of Learning Mathematics* p. 69.

Having what Hilary Putnam calls a "full-blown" concept "chair" involves what Frege speaks of as the kind of complete knowledge of it that can distinguish true and false, correct and incorrect uses of the word in every case. Manifestly, the more technical or sophisticated a term is the less we attain to this ideal. Most of us would score highly were chair-like words the entirety of a language, but at the other end of the scale we would often score very poorly over usage of the word "true" (unless we arbitrarily limited its scope). Formalisation and axiomatisation is one means we employ to overcome this deficiency, this lack of full-blown-ness. Moreover, this full-blown-ness is not restricted to correct usage according only to the *internal* criteria of a language or formal system; everyday practice quite rightly insists that to qualify as *understanding* a word (having a concept) we should also display appropriate usage vis-à-vis the real external world.

Historically the delineation of the basis of mathematics occurred relatively late. Euclid did not invent plane geometry; his genius lay in gathering together a plethora of results and deriving them from simple principles which he also chose. A rigorous justification of the differential and integral calculus, known since Newton and Leibniz, has only been possible in recent times. The basis of arithmetic remained unknown or incorrectly formulated until Frege attacked the scandal of its woolliness in the nineteenth century. The curious spectacle of mathematics justifying itself rigorously only long after its results have been accepted and applied indicates that mathematics is not based upon its "foundations" at all; indeed, as we have seen, the word is misleading, for foundations must precede superstructure, and manifestly they have not done so. Of course, if we accept the formal — non-formal distinction advocated in this book, the problem is easily resolved by saying that whereas these *are* the foundations, they have only recently been formalised as such, and were previously tacit. But the process nevertheless resembles the sight of an untidy researcher leaving behind him a trail of chaotic and poorly-documented results until the day when his study became so disorganised that it became necessary to clear things up. This is what has happened during the last century, because until then *alternative* systems of logic and arithmetic were either not known or not studied; the absoluteness of Newton's cosmology was paralleled by a misleading straightforwardness in mathematics. But while Lobachevsky and Riemann were developing alternative geometries, other mathematicians were busy speculating about alternative logics which arose from altering some of the axioms of "classical" logic.

Conversion involves change; change involves risk; risk must be balanced by prospect of rewards which justify it. I failed to learn French because my fear of error prevented me from exploring my own uncertainties; many mathematics students fail because they are intimidated by either the subject or its teachers; many people reject Christianity or avoid it because they are afraid of institutions, clergy, and its kill-joy image, because they see it not as a way to achieve true freedom, but as a trick designed to persuade them to surrender it. Therefore conversion will only occur when the boundaries of our play-space are secured by those competent to enable us to explore both the inadequacies of our present way of life and the opportunities available to us in others. That the mainstream churches fail to inspire trust, love, or interest is therefore a major stumbling-block to the acceptance of the Gospel of Jesus Christ.

These discussions of circularity, conversation, and conversion enable us, by taking account of the Fallacy of Democracy and the inevitability of circularity, to formulate principles to guide us in our search for truth:

(a) mutually incompatible systems cannot be bridged logically, and therefore logical conversion from one position to another is not possible;
(b) we must all occupy or espouse some system of beliefs or axioms, even if it is only that of the culture into which we are born, and which we consequently chose by default;
(c) the system we adopt deliberately or by default may be mistaken, but that is inevitable and therefore only a relative deficiency;
(d) to choose a system is to choose it as true and to reject other systems as false, despite the fact that it may conceivably be false;
(e) to believe that a system of ideas or beliefs is true is to be obliged to try to convince others of its truth if we are not to embrace a version of relativism;
(f) the theological imperative to be discussed in the next chapter indicates not only that it is wrong to try to force others to adopt our chosen system, but that it is counter-productive as well;
(g) to encourage others to adopt a point of view we share we must abandon the attitude that we are right and they are wrong, which they will experience as condemnation, and adopt the view that we share a vision which they could enjoy, provided we do not behave in a way which alienates or intimidates them; in other words we are not engaged in coercing others into admissions of failure, but in sharing a gift which is greater than us all.

A Gödelian Cardinal Principle would be:

Cardinal Principle Omega — For some situations there are no Cardinal Principles.

MATHEMATICS AND REALITY

The effect of axiomatisation was to wrench mathematics away from applied science and to enable it to develop as an autonomous discipline studying structures which seemed to be of interest without regard to their usefulness. But that makes the question of mathematical truth problematical, because "true" no longer means "true of the physical world", since mathematicians are free to invent "true" theories which do not pretend to bear upon or refer to the physical world. The question then arises how we decide which of the many systems of mathematics we have available should be used in our applied science, for example in cosmology.

In one of his finest and most famous essays, Einstein expressed the matter thus:

> These axioms are free creations of the human mind. All other propositions of geometry are logical inferences from the axioms. ... The axioms *define* the objects of which geometry treats. ... This view of axioms, advocated by modern axiomatics, purges mathematics of all extraneous elements ... but such an expurgated exposition of mathematics makes it also evident that mathematics as such cannot predicate anything of objects of our intuition or real objects.
>
> Geometry and Experience.

He goes on to say that we set up our mathematics and then *add* one further axiom or assertion, namely that the universe is disposed according to this particular system of geometry.

In modern mathematics, which might appropriately be dubbed *post-axiomatisation mathematics*, we can start from a collection of axioms, alter one or more as we have seen, and then see what happens, i.e. which of the old results remain true, or can be derived in novel ways. Such procedures highlight the problem of how concepts are formed when they are artificial, that is when the results they give rise to and reflect cannot be intuited other than by building up a series of theorems and performing what Polanyi calls *tacit integrations* of them as particulars in order to come to some kind of focal awareness of their meaning. Polanyi calls this *the domain of sophistication* where our formal expressions lead us to realms of meaning we cannot intuit, as for example when we extrapolate geometry from three to four or more dimensions. What does a four-dimensional cube "look like"? There is no answer, because "looking" is an ability developed in three or less dimensions which we seem to have no means of extending to four. (There is a fascinating account of one way of overcoming this difficulty in Davis and Hersch, *The Mathematical Experience* where they discuss computer models of a four-dimensional cube and how they enable us to develop a "feel" for them.)

Mathematics has now brought us to a kind of problem which many would discount as "mysticism". The most logical and rational of disciplines, identified with the utmost *reality*, itself raises the question of the nature of reality. We think we know what a point, line, square and cube are. Nothing could be simpler. But what is a four-dimensional cube? We cannot "see" one, build one, draw one or imagine one, except in some kind of projection. But we *can* say how many faces, edges, and vertices it has, just as we can for cubes with five, six, seven or n dimensions (and the interconnections of the latest generation of supercomputers are designed on the pattern of an n-dimensional cube). Is this meaningless nonsense, a harmless game, or an indication that our concept of reality is defective?

In addition, mathematics deals with triangles, squares, circles, lines and planes which nobody has ever seen, and nobody can construct. We *think* that we can imagine a circle because we know a precise analytical definition of what a circle is, and because we have seen plenty of shapes approximating to circles with varying degrees of success. But can we imagine a *pure* circle? Need we even try? Frege showed that words do not name mental images, and that understanding a word does not involve or require a mental image.

SCIENCE AND REALITY

The problem, to approach it from another direction, is that we tend to identify *physicality* with *reality*. Our sense-experience tends to legislate for what we are prepared to regard as real. A more general problem arises from scientific accounts of the way things are, which led John Locke to distinguish between primary and secondary qualities.

> Primary qualities like shape, size, number and motion have been treated very differently by physicists, at least since the seventeenth century, from secondary qualities like colours, sounds, and tastes.
> J. L. Mackie, *Problems from Locke* chapter 1.

The solid table I write on is not solid; my roses are not coloured; these impressions are functions of my particular kind of physical existence and sensory apparatus. This is all very alarming, because it seems to suggest that my senses are only useful for getting me through the day, and not for telling me how the universe really is. The atoms making up my table are, relative to their own size, separated by oceans of empty space; my roses look yellow only because the wavelength of the light they emit as electrons drop back into more stable orbits makes a particular kind of impression upon my rods and cones, which in turn excites my optic nerve and generates an impression I call "yellow". What is happening? Is the table solid or not; are my roses yellow or not? Is my whole life an unreal dream constructed for convenience but bearing no resemblance to things as they are? Nothing seems to be simple any more. And of course we do actually *need* things to be simple. Our friend the man in the street finds life complicated enough with paying his bills, raising his children, mending his car, and doing his job. He doesn't need things to be made even more difficult by people with nothing better to do than to point out to him that even his sense-impressions are leading him a merry dance.

Is there a way out of this apparent impasse? We certainly do not wish to dispense with science, or to advocate a philosophy incompatible with its findings, but it seems as if by incorporating science into our world-view we are forced to deny the reality of things we all take to be self-evidently true, such as the solidity of tables and the yellowness of roses. Why do we regard a description of the world in terms of atoms and waves as *more real than* or *superior to* descriptions in terms of solidity and colour; and what kind of understanding of reality is entailed in regarding solidity-words and colour-words as keys to it? On what basis, given the insecurity of our belief in the yellowness of roses, are we so confident of the truths which "as children of our culture" we never question, and so dismissive of the claims of other cultures and other times which we never consider?

The man in the street avoids such worries by *restricting the kinds of questions which he asks*. The rejection of questions is not always illegitimate: life would be intolerable if we were perpetually forced to reconsider elementary assumptions, or major problems beyond our comprehension. But rejection of questions we find uncomfortable is a primary device we use to defend ourselves against change and to preserve the illusion of the adequacy of existing systems of ideas.

The need for change if we are to grow together and towards the truth, and the need to limit the number of questions we ask, raise the question of how we know which questions to permit. This is the same question we encountered in another form when we observed that mathematics does not elaborate every conceivable theorem derivable from a set of axioms, and that science does not pursue every conceivable experiment or hypothesis which could arise from a given set of assumptions. Mathematicians and scientists make choices about which problems they should tackle. For Polanyi, the ability to choose good problems is a key skill in any scientist, and it may also be seen to be a necessary skill in a mathematician. (Gödel's greatness, for example, lay as much in perceiving the possibility of an incompleteness theorem as in articulating and proving it.) Hence the specialised skills of scientists and mathematicians find their counterparts in the everyday need to decide which problems require solution or which questions should be allowed to disturb our existing world-view. And the question "does it matter?" in turn puts our systems of value under the spotlight.

I am suggesting that *the question about legitimate questions* (a meta-question, if you like) is just such a key problem, since it opens out into an issue of the widest possible significance not merely as an intellectual exercise, but in the face of the increasing polarisation of our world, which leads to vast expenditure on armaments, desperate dangers for the human race as a whole, and religious schisms which threaten the security of innocent people through bomb outrages and hijackings. These rest upon our refusal to allow our firmly held beliefs and convictions to be called into question, in other words upon the belief that we are right and everyone else wrong.

We have fixed views about what is real, and because those views are based upon our conviction that we know what we know (itself a facet of the self-confidence of our scientific and mathematical culture), we tend also to imagine that we know what we *need* to know, and *a fortiori* what questions we need to ask and answer. In conjunction with our need for certainty and security these features of our world-view prevent us from taking seriously any questions which might unsettle its stability or our confidence in it. To become more human we need to be less defensive. One way to become less defensive is to demand less proof of the things we find strange, and more proof of the things we take for granted, to take more on trust by opening our minds to the possibilities arising from alternative world-views. But can we relinquish our demands for proof without relinquishing our hold on truth as well?

PROOF AND REASON

Once proof is understood along the lines given here, and once the force of formalism is dissipated, we can see it as a servant of human enquiry rather than a despot ruling tyrannically over it. We will be happy when we can find convincing proofs because they help us to establish that we have not lost our way, but we will not be dismayed if we do not find them. There is a pertinent analogy with new drugs in the treatment of terminal disease. Doctors simply *do not know* what the side-effects are in the long-term; they have no proof that the drug is "safe" (for, in fact, no drug is "safe"); all that they know is that the need of the patient is desperate, and that the outcome of doing nothing is absolutely certain. The presence of the possibility of error is then less important that the availability of a slender hope. Now in most situations our decisions are less dramatic, but we would be foolish to deny ourselves a field rich in its potential for knowledge merely because it *might not be safe*. In the last analysis, of course, life is not safe; neither is it intended to be.

Proof is primarily a confirmation that things are going well. It does not, and logically cannot supply specifications for what "things" ought to be tried that might or might not go well. The hypothesis gives rise to the search for the proof, not the other way round. But what gives rise to the hypothesis? Really the *nose* for a good problem, a rich field of enquiry, which encourages us to dwell in it in the hope of reaching sufficient understanding to perform the tacit integrations which will generate genuinely important new problems and results. But "nose" is a human quality which embraces the whole panorama of human faculties from dispassionate reason to feeling, emotion and belief. It is therefore to reason, feeling, emotion and belief that we now turn.

REINTEGRATING REASON

IF the existence of God were demonstrable by mathematical proof, or his nature discernible by application of the scientific method, the self-sufficiency of man's rationality would be confirmed and reinforced. In fact, because neither his existence nor his nature can be so determined, God's being-for-us demands that we abandon that centre which is within ourselves (both as regards reason and being) and find instead that true centre which is in the Other. We must be prepared to lose ourselves in order to find ourselves again. In other words, the claims of Christian theology stand as a challenge to and contradiction of the self-sufficiency of unaided human reason, and as a demand that we relocate our centre beyond ourselves in the other. Theology demands a reformation of our concept of reason.

This *theological imperative* involves Christians in a stance which their culture is likely to find offensive: because knowledge of God demands that we locate our centre and orientate our reason with reference to that which is other than ourselves, the same obligation is laid upon us in all our enquiring and knowing. Insofar as God's being-for-us is also *personal*, this is not a matter of searching for impersonal objective criteria, but of discovering a new basis for knowledge.

Polanyi proposes a model of knowing based upon a dynamic conversation between man and his environment in which we are both affirmed and called into question. This mode of knowing leads us to personal knowledge: "personal" in that it involves the knower and the known in a relationship of fundamental faithfulness in which neither is forced to be or to become what it is not; "knowledge" in that out of this relationship of faithfulness there arises a depth of understanding which leads us to state our findings with universal intent.

Personal knowledge involves a distinction between *self-centred* reason (reason centred upon its prior assumptions about itself, that is reason centred upon itself and as such essentially unrevisable and static), and *other-centred* reason (reason which is open to revision by virtue of demands made upon it by the objective other, where "other" is chosen deliberately to include the animate, the inanimate and even the divine).

Self-centred reason is dominated by fear of error, and as such by fear for self. It reflects a concern not to be wrong, but limits the extent to which we can be right. Because it usually remains blind to the inconsistencies at its heart by not perceiving its hidden assumptions it tends to adopt a sceptical attitude which is fashionable (unless it remains sceptical about everything), and is easily persuaded to be sceptical only about the things which "the crowd" is sceptical about. As such it reflects a concern to be in step, and is often characterised by a witty and destructive scepticism aimed at claims which do not satisfy its own rigidly circular criteria. It tends to be mistrustful of the emotions, and can lead to a destructive division between reason and emotion. At its worst it is negative about life, and rejoices in sucking others into its cynicism; at its least harmful it simply infects people with inability to hope (since hope is based upon something which self-centred reason cannot analyse).

Self-centred reason is incapable of trust, because there can be no water-tight means of eliminating doubt and establishing a basis for trust; it cannot even trust its own emotions as companions or guides, and so it leads to a divided and fragmentary life which can find enjoyment only in wielding the destructive tool of scepticism against the supposedly irrational beliefs of others. It is prey to the prejudice that its own age and its own wisdom are superior to all other ages with their wisdom; therefore neither the world nor the past has the power to call it into question, and insights of past generations and other contemporary human beings are lost to it. Believing in the absence of assumptions in its own stance it is easily led into self-righteousness in which it puts questions to the world but is never called into question by the world.

Bertrand Russell once addressed to God the question "why have you made the evidence for yourself so insufficient?". What does the inaccessibility of the existence of God to verification by mathematical proof or scientific method tell us about the nature of God? What does the fragility of the evidence for Christianity (as measured by the criteria of self-centred reason) tell us about the Christian concepts of reason and truth?

RELIGION

It is intrinsic to the nature of God that he is other than Man or the universe. Any conception of God for which this is not true uses the term "God" in a dishonest and misleading manner. But much so-called theology is prepared to dispense with this otherness today, and to replace God with a human creation, a mere psychological projection.

> [Do you believe] that God is transcendent to human affairs and to human attitudes ... that God would exist whether human beings and their attitudes existed or not? ... And I think it worth asking oneself very carefully when confronted with some reinterpretation of Christian doctrine whether it passes this test: that it represents God as a being who would be there even if no human beings ... were there. If it does not, then I suspect you no longer have any form of Christianity, but probably some form of religious humanism.
> Bernard Williams, "Has 'God' a meaning?", Question 1, p. 53.

Whereas it is necessary for any conception of God to pass this test, it is by no means sufficient. Spinoza's God passes it by virtue of being indistinguishable from Nature, and as such independent of human existence; but many modern "gods" do not. Theologians, recoiling from the ferocity of sceptical questioning have adopted what Hans Albert calls *immunisation strategies*, theologies which preserve the word God but with only a shadow of its full-blown meaning.

> Committed atheists are understandably irritated when theologians defend a completely vague, overstretched, *eroded concept of God* and blur the frontiers between belief in God and atheism.
>
> Küng, *Does God Exist?* p. 334.

The *idea* of God edifies, the *possibility* of God fascinates the intellect, and the *superstition* of God bestows power on the unscrupulous, but the *utter reality* of God offends and terrifies, even as an idea. Yet just this terror produces a huge market for those who can weave conceptions of God that immunise us against his reality. We displace belief in the Lord of heaven and earth onto belief in a tame human creation controllable by human reason.

Thus the distinction between self-centred reason and other-centred reason cuts across stereotyped distinctions between so-called religious and unreligious people. Bonhoeffer was right to be suspicious about the value of the religious *a priori* for the Christian faith. It raises the question whether a particular instance of religious faith is based upon the self or upon the other, whether it is sustained to satisfy a self-centred craving, or is a free response to the perception of truth which lies in the other. It asks whether religion is or is not, after all, unbelief, that practice which reinforces our sense of the adequacy of the centre we find in ourselves by making us feel whole and self-contained, that process of self-sanctification.

Conversations with Jews, Moslems, Hindus and Buddhists confirm the appropriateness of this demarcation. Religious faiths and denominations draw their membership from both categories, and most people fall at one time or another into each. Religious institutions are torn between the need to supply what self-centred people demand, and their responsibility to encourage other-centredness. The distinction is between those for whom religion is essentially a means of remaining the same, and those for whom it is a means of changing. The same distinction appears to apply to the "unreligious" as well, which suggests that the lines of demarcation between those with a rationality fundamentally open to the possibility of the true God, and those fundamentally closed to it, should be drawn rather differently from those which merely separate the religious from the unreligious.

HUMILITY

What we have called the self-contradiction of post-Enlightenment thought lies in its double emphasis upon the individuality and autonomy of man (that is the right of each individual to make up his own mind according to the principles of self-centred reason), and the objectivity and impersonality of science (that because what we know is based upon methods of reason which are without presuppositions, the conclusions are objective in the sense of not being the responsibility of those who discover or affirm them). This split licences subjectivity under the guise of objectivity: "I do as I must do according to the objective criteria of reason, and whatever ensues as a result is no fault of mine." The perversion of Augustine's aphorism "love God and do as you like" is "love reason and do as you please". The seeds of the extraordinary contradictions apparent in modern society, in which the massive success of science based upon application of rational processes to the investigation of the nature of the universe is in sharp contrast to the increase in interest in astrology and the occult based upon the freedom of man to do and believe as he likes according to the dictates of his own conscience, can be seen to lie deep within the hidden presuppositions of critical philosophy.

Self-centred rationality will often be characterised by a mixture of empiricism and utilitarianism, reliance upon evidence and usefulness. The reliance upon evidence is consistent with its refusal to take risks: it prefers to point to evidence as an impersonal justification of its position (and as such a justification which does not involve responsibility), without acknowledging that evidence must be interpreted in order to be intelligible (there are no "raw facts"). It will rely upon perceived usefulness because it has no room for imagination (since imagination is orientated towards the future, and self-centred reason is rightly doubtful about the applicability of its tools to the future). It will therefore add to its armoury of destructive weapons the powerful question "what use is that?", while simultaneously defining what constitutes an adequate reply. The only honest answer we can give in many circumstances is "in your terms (i.e. the terms of self-centred rationality), no use at all".

Other-centred rationality recognises the priority of that which is other than ourselves, and leads to the insight that it is the other which calls us into being and enables our becoming. It argues that to cut ourselves off from the other by assuming a centre within ourselves is to cut ourselves off from the source of life; it is to die. But this is not the reason for its other-centredness. Whether in science or history, religion or everyday life, the other has the power to evoke our love. It is more precious than the self, and commands respect for itself as it is without violating our selves as they are. It addresses us by being there for us; it addresses me by being there for me. It is of the essence of this love that we feel moved to speak of the other despite our recognition that as products of our feeble minds any words we utter will be inadequate properly to describe it. In saying this we are not speaking only of the Other which is God; we recognise that our words and concepts are as inadequate to describe anything, even ourselves. Nevertheless, we do speak, because to speak inadequately is better than to remain silent, as Augustine wrote in his *De Trinitate*. This remains true in science and history, religion and everyday life. We cannot grasp the other, but we can communicate impressions. Such knowledge follows closely Hilary's distinction between *comprehension* (complete knowledge which grasps the whole essence of the known), and *apprehension* (as that which recognises the limitations of knowledge and is content to know what it can).

In cosmology, for example, we find that our mechanics, even with the added complications of the theories of relativity, only begins to purchase upon the reality of planetary motion. We cannot yet solve the three-body problem in Newtonian mechanics, or the two-body problem in Relativity Theory; what makes us so confident that we can comprehend the universe? Reality is much more rich than anything we can model with our best science. Only a concupiscent, possessive attitude refuses to recognise the fact, and pretends that it can comprehend the universe in a set of equations, for when we have said all that can be said in our equations we have left unsaid most that is worth saying, and unseen most that it worth perceiving. Scientists are far more circumspect about their achievements than the layman who likes to believe that scientists will one day know everything and solve all human problems.

REASON AND VALUE

Just as the distinction between self-centred and other-centred reason cuts across traditional divisions among the religious and unreligious, so it challenges disciplines such as mathematics and the natural sciences, history and philosophy. Any discipline which wishes to be true to its subject-matter, and not to limit itself by prior conceptions about what is or is not possible and reasonable, must be other-centred. The discipline which seems most likely to become self-centred is mathematics, since mathematics seems to investigate the properties of concepts which are essentially internal to itself. In mathematics we have strict criteria of faithfulness to the subject-matter in rules of inference which determine whether or not acceptable proofs have been supplied for the hypotheses advanced. But to leave the matter there would be to miss a feature of mathematical research which throws light on research in all other fields, because the rules of inference in mathematics are so clearly defined and rigid. The fact that a mathematical result is proved does not mean that it, or the theory of which it forms a part, has any *value*. But "value" should not be taken as synonymous with "utility". A result or theory in mathematics is deemed valuable not on the basis of anticipated utility (will it make rocketry easier, for example), but on the basis of the existing problems it resolves, and the suggestions it makes about new avenues of enquiry. In other words, an appreciation of mathematical value involves the ability to perceive as-yet-unknown possibilities which may arise from a result or hypothesis. But, as we showed in the previous chapter, such insight into the implications of a result does not arise merely from familiarity with it as a formal statement; it arises from insight into the mathematical reality of which that statement forms a part. The component of reason which this illuminates is not the logical deductive component, important as that is, but the visionary component, the component of reason which yields intimations of fruitfulness, which shows us whether or not to pursue a line of enquiry further. An attitude to reason which claims that the only things we should countenance are the things which satisfy the strictest criteria imposed by methodical doubt leaves out of account the component of reason which helps us determine which results to argue for, that is, which of the countless possible candidates for our adherence we should even begin to examine. And without some such selection criterion the principle of methodical doubt is useless.

Hume showed that the principles of methodical doubt destroy science by destroying the theory of induction whereby scientists infer from the past conditions which may obtain in the future. There is no necessary connection between logic and experiment. No theory can therefore be justified according to inductive principles, since the best that we can say is that, for example, on every occasion when we have put our theory to the test in the past it has proved correct; that is absolutely no guarantee that it will prove correct for ever. But the fact is that we need to make predictions about the future, and therefore we allow a practical necessity to over-rule our scepticism. We also need some kind of answer to the question "what am I to do with my life?" How do we make such choices? Using largely the same feature of reason that we use in choosing promising mathematical ideas to pursue: we identify possibilities and follow those which seem to be congenial. The difficulty is that the kinds of people we think we are and the kinds of people we think we would like to be are all limited by the kinds of people we think it is *possible* to be, which is limited by what we think to be reasonable.

MEASUREMENT

I remember from my primary school a large, thin, hardcover book entitled *Man Must Measure*. It chronicled man's measuring from earliest civilisations through to the twentieth century. The title, however valuable or accurate the contents, which I have forgotten, tells a story, for it is a deeply ingrained characteristic of human beings that they must have numerical equivalents for their experiences. Socrates observes in Plato's Republic (VII:522d) how foolish Agamemnon was made to look when Palamides claimed to have invented number, implying that Agamemnon had never counted his troops. We all feel more secure when we can give precise answers to questions such as "how much?", and "how big?" and "how many?", and we have contrived to order society in accordance with amounts of money, acres of land, numbers of qualifications, so that measurement, like science itself, has moved from being only our servant to being in some respects our master (or at least a dictator of our system of values). When we are denied accurate measurement, we feel insecure, vulnerable to error, and moved (sometimes) to invent spurious scales of measurement. "Is he intelligent?" becomes "what is his I.Q.?"; "Is he educated?" becomes "how many examinations has he passed?"; "Is he successful?" becomes "how much money does he have?" or "how expensive a car/house/holiday can he afford?" The invention and development of computers has accelerated the trend towards conversion of information into digits. Berman has coined the unhappy term "cybernetization" to describe the process whereby human values are made amenable to computer technology. But suppose measurement forces us to distort the world before we can count it by generating a number-based world-view that does violence to the way the world is? Suppose that perfectionism inherent in measurement and mathematics actually generates false conceptualisations of the world in terms of a number system that is not wholly appropriate? Suppose, that is, that numerical perfectionism leads once again to inversion of our understanding of the thing measured? That this was the case would be obscured because the inversion itself would lead us to ask the kinds of question that require a numerical answer. The system of values governing our minds would become self-authenticating by permitting us to ask only the kinds of questions which can be answered within the system. Such processes resemble Protection Rackets; that is, they present themselves as friendly solutions to a severe problem of which they have themselves been the cause.

Anyone who has attempted to advocate a value-system which is not measurable will have experienced the intense opposition, even hatred that such a suggestion evokes. The reason for this is not difficult to find. Most of us have sold out in some respect to the "man must measure" syndrome; we each have our place on one or more ladders whose rungs are marked in pounds or dollars, intelligence quotients, examination passes, cups or prizes. We locate ourselves with respect to these measurements, finding it reassuring that although there are many above us there are (we hope) even more below. As self-centred men and women we have an enormous investment in suppressing questions about real values, which defy measurement according to the "objective" criteria which force the results upon everyone. With an I.Q. of 135 he must be intelligent; with a large bank balance he must be successful; with a lot of qualifications he must be educated. But these conclusions follow only if we accept the premise that what we can measure is what matters, that the measurement is the value; only if we accept the protection proffered by the racketeers. (I was intrigued to hear a psychologist saying the other day that the reason why "Game Shows" are so popular is that a high proportion of the public like to believe that the important questions are the ones with straightforward answers.)

The observation that reality may be less amenable to accurate measurement than we like to think is not new, and a long-standing solution, partly reinforced by the principle that the measurement is the value (we even call our measurements "values"!) is to invent a new reality (such as money) which we know to be measurable. Money is unusual in that unlike most objects in the universe, one pound or dollar is exactly like any other, so that saying "I have one hundred pounds" destroys no information. Nobody would ask "what is each of your one hundred pounds like?"; but if I say "I have one hundred leaves from an oak tree" I have overlooked a vast amount of information which would justify the question "what is each like?" if anyone were interested (which I think they should be, odd as it may sound).

> [The fallacy of misplaced concreteness] consists in neglecting the degree of abstraction involved when an actual entity is considered merely so far as it exemplifies certain categories of thought. There are aspects of actualities which are simply ignored so long as we restrict thought to these categories. ... Philosophy has been misled by the example of mathematics; and even in mathematics the statement of the ultimate logical principles is beset with difficulties, as yet insuperable.
>
> Whitehead, *Process and Reality*, pp. 7ff.

We countenance the destruction of information because we cannot handle it (computer scientists call data they cannot put into the machine "dirty data"!), and because we do not wish to acknowledge the implications of that inability. We would like to think that when we have counted and measured we have comprehended, and that nothing is left unsaid that is worth saying; therefore we refuse to admit that we may have left unnoticed the most important features of the objects under scrutiny (the uniqueness of each individual leaf, for example). We rationalise our disguise of apprehension as comprehension either by defining knowledge as that which can be known precisely or by an appeal to utility: in theory what you say is correct, but in practice it doesn't matter. Cumulatively of course, it does matter, because by eliminating the aspects of the world which do not fit our theories (either because they are too diverse or because they are too complicated) we are misled into conceiving of the world solely in terms of our theories. We forget the "remainder term". We literally recreate the world in a measurable form.

LANGUAGE AND REALITY

The process of "cybernetization" is much older than the computer; it is as old as science, as old as theory, and therefore as old as language. Nevertheless, with language in all its vagueness we are able to rely upon corrections and interpretations supplied by our understanding to render our usage appropriate in different contexts. Nobody imagines that in two contexts A and B "this is a table" states that the table in context A is identical to the table in context B; but we can only avoid this mistake by virtue of the richness of our conceptualisation, not by a mere analysis of the formal terms.

Frege battled throughout his life against idealism and psychologism. In particular he rejected the view that words name mental images, and that understanding a word involves having a mental image. Understanding the word "table", for example, would be impossible if all we had was a mental image of a table: how would we recognise strange tables, or the word "table" in an unfamiliar context? Understanding a sentence is not an empirical matter; it is not necessary to be able to *point to* some event or object in order to understand "Napoleon was a great general"; indeed, it is impossible to do so, for there is no such event or object "out there" in the physical world.

> ... possessing the full-blown concept is not a matter of possessing *further* images (say, images of sentences, or even of whole

discourses), since one could possess any system of images you please and not possess the *ability* to use sentences in situationally appropriate ways.

Putnam, op. cit., p. 5.

Nevertheless, we do have mental images, so if they are not the means of understanding words, what are they?

Kant understood the deficiencies of a theory of language which identified words with mental images, but he made little of the insight:

> No image could ever be adequate to the concept of a triangle in general. It would never attain that universality of the concept which renders it valid of all triangles, whether right-angled, obtuse-angled, or acute-angled ...The schema of the triangle can exist nowhere but in thought. ...Still less is an object of experience or its image ever adequate to the empirical concept ...The concept "dog" signifies a rule according to which my imagination can delineate the figure of a four-footed animal in a general manner, without limitation to any single determinate figure such as experience, or any possible image that I can represent *in concreto*, actually presents.
>
> *Critique of Pure Reason*, A141.

A mental image (of a circle or face, for example) is as much a name as a word; that is, it stands for or represents one particular aspect of a concept; it is a thought. Frege believed that only a sentence represents a thought, and that we only have thoughts by having sentences occur to us. My question would be whether, for example, our mental imagery does not in fact represent an aspect of our understanding in exactly the way a sentence does? Our mental images are less photographic than we think: imagine a place you know well; now try to count the panes of glass in the house; can you do it? Look at a portrait of a familiar figure; does it look exactly like him? Usually not — we do not expect portraits to look exactly like the subject. A mental image is a particular kind of formalisation of our concepts, our non-formal world. If we deny this we are unable to account for the brilliance of a great portrait painter who can express on canvas his insight into the nature of the person he is painting in a way quite beyond any photograph. Great artists are valued for more than their technique: we value their ability to conceive of their subjects in ways which expand our understanding; they show us more than we can see.

DECISION AND CHOICE

According to the criteria of other-centred reason, whereas the rules and procedures of mathematics and science remain normative wherever proofs or verifications are to be found, the primacy of the centre outside oneself always retains the power to over-rule the requirements of logic and science by presenting us with some intimation which reveals possibilities more important than mere avoidance of error or danger. It does this by promising us access to a richness of meaning and possibility which embraces our imagination, evokes our love and induces in us a proper humility about the criteria we more usually employ. This experience of the self-disclosure of the other is the central motivation for all discovery in all fields; it is the source of the idea, hypothesis, theorem which although new, untried and unproven grasps our imagination and demands of us that we follow, forsaking our nets and the safety of our boat.

> Instinct, intuition, or insight is what first leads to the beliefs which subsequent reason confirms or confutes; but the confirmation, where it is possible, consists, in the last analysis, of agreement with other beliefs no less instinctive. Reason is a harmonizing, controlling force rather than a creative one. Even in the most purely logical realm, it is insight that first arrives at what is new.
> Russell, "Mysticism and Logic" in *Mysticism and Logic*, p. 19.

The feeling that there is something out there more important than security, and more rewarding than certainty is far less tangible than, for example, a "Damascus Road" experience. Polanyi described it in terms of a heuristic field whose lines of force draw us away from the familiar towards the integrative insights and further expansive hypotheses and conceptualisations which are life, and light, and joy, and peace. This evocation, far from being general, an address to the whole of mankind, is likely to be experienced as specific and personal; an address to me.

In the chapter on proof I examined some of the ways we explore conceptual and heuristic fields. Here I would like to use a different and dangerous example drawn from the way we choose marriage partners for life. (It is obviously a dangerous example because of the number of such choices which prove wrong.) My eldest daughter, then three-and-a-half, suddenly asked at lunch one day, "what does 'marry' mean?" After the usual sequence of inadequate parental answers my wife changed tack and tried the flippant, "well, Mummy looked at all the men in the world and then chose Daddy"! And, of course, the reason why that is funny is that it is absurd. Not, that is, simply because it is untrue, or because it would be impossible to perform such a task, but because *nothing of the kind* is the case. One simply does not choose a partner in such an empirical way (any more than one chooses a mathematical theorem by trying every combination of known symbols). Falling in love involves, or should involve, responding to the discovery of the other in whom one can find one's centre; it involves an unpredictable, and therefore properly awe-inspiring and frightening leap into a promising unknown full of as-yet-inconceivable possibilities, and to the risk of making the famous, much-maligned, inconceivably "irrational" but contextually appropriate promise, "for better for worse, for richer for poorer, in sickness and in health, to love and to cherish, till death us do part". And the same thing, in a way, is true of mathematics; and the same, in a way, is true of science; and the same, in the most profound way of all, is true of God.

REASON, FEELING AND EMOTION.

Integration of reason, feeling and emotion is especially important because of the influence of *feedforward* systems in our personal development, the ways in which our future attitudes and abilities are shaped by current experiences which are to some extent within our control. Let us begin with an illustration. We are all acquainted with people who are described as "accident-prone", and we charitably grant them a certain licence to be clumsy because we see no way in which this apparent handicap can be altered or removed. However, we often find that such people cease to be clumsy when engaged upon a task which matters to them such as some sporting activity; and we have no misgivings about associating their physical clumsiness with other aspects of their behaviour which tell us about their attitude to the world, which may include absence of observation, obliviousness to their surroundings, and a general mental state which does not notice the world with what we (perhaps in a prejudiced way) regard as normal sensitivity. The same is true of those who are excessively tidy or untidy, punctual or unpunctual, sensitive or insensitive, and so forth. In fact, the same is true of everyone in some respect or other. Aspects of our behaviour tell our fellows about aspects of our world-orientation (which may be regarded as equivalent to our mind). Behaviourist reductions of this regard such observations as justifying assertions to the effect that therefore we are all "merely" or "nothing but" products of our genetic and environmental heritage (as in Skinner's theories), or that the workings of our bodies are the same thing as the workings of our minds (as is argued in Ryle's *Concept of Mind*). Such interpretations are concerned to understand all things from the bottom up. Contrast the no-less-legitimate interpretation which sees in this rich diversity of human dispositions our strength and greatness, and in our capacity to be shaped by certain environmental factors a real hope for our future. For the fact is that by virtue of our ability to choose the environment we live in we can govern the kinds of people we will become. To immerse ourselves in science or literature, pleasure-seeking or sport is to alter ourselves for the future, and therefore to act selectively in deciding and determining the personality we develop. Our motivation to be for or against some line of reasoning is not the mere product of blind chance, but a product of a lifetime of deliberate and intuitive selection in which we have placed ourselves in positions which will sensitise us to certain kinds of stimuli and desensitise us to others. Such considerations enable us to assert *both* that we are influenced by our environment and genetic inheritance *and* that we are responsible for the kinds of people that we become. Accident-prone people are not to be exempted automatically from responsibility for their actions.

Passive ideas of the way conditioning works, largely formulated by analysing the influence of the past upon the present, must now be augmented using feedforward concepts to include the possibility that we can consciously and deliberately make choices about what we will become by tailoring our present environment. Our laws and social practices commonly include allowances for disadvantageous early conditioning, but exclude consequences of deliberately chosen feedforward influences (as the example above illustrates). Yet the cynical German aphorism *"Man ist was er isst"* (man is what he eats) is true to the extent that our minds can only chew over the cud of formerly absorbed food (which points directly to one of the central meanings of the Christian Eucharist). If that food is savage, punitive, simplistic or pornographic, then the mind cannot but regard such attitudes and practices as normative.

The problem that rationalism has with feedforward systems is that nobody can foresee all the consequences of current influences, any more than we can foresee all the theorems which may arise from a set of axioms in mathematics. Therefore the data necessary for a rational assessment of which influences we should subject ourselves to are inaccessible and unknowable. But we can decide if we allow our total personal response to possibilities as they manifest themselves to us to be our guide.

A similar problem concerning foresight and predictability lies behind a common misunderstanding in computer programming which leads to the assertion that a computer can only do what you tell it to do.

> [If this common slogan] is taken to mean that everything the computer does is done at the behest of the instructions in the program the slogan is, of course, true. But if it is taken to mean either that the programmer can foresee everything that the program will do, or that the program will do *all and only* what the programmer intended it to do, then it is false.
> Boden, *Artificial Intelligence and Natural Man*, p. 7.

Rationalism also tends to suppose that we always understand why we do things, and why the stratagems we adopt work. This is not the case. We constantly engage in behaviour designed to achieve a given end despite our ignorance of how it does so. We do not understand how our brain implements decisions, but we constantly and inescapably rely upon it to do so. Similarly, the only convenient way we can talk about emotions, feelings and beliefs is using "mind-talk" since the "brain-talk" is not available. But the legitimacy of this is confirmed every day. We adopt strategies to control our emotions without knowing why they work, and without supposing that there is a clear causal link between the strategy and its effect. For example, "counting to ten", "having a cup of tea", and "taking it out on the garden" are common devices we employ to disperse anger, anxiety and grief. Many people find counselling and psychotherapy beneficial, or simply "talking it over with someone", but nobody understands how such activities benefit us in neurological terms. What matters is that they *work*: a distressing or insoluble problem has been resolved.

Consider another example drawn from medicine. There is always a great deal of argument between counsellors and doctors about the relative efficacy of talking and prescribing medicines. The debate often sounds as if the connectedness of brain and mind-states had not been appreciated. But there is no reason to suppose that talking something over with someone must necessarily have a different effect on the brain from prescribing a drug. The advantages of the first are that it might be more specific, has fewer side-effects, and may be permanent, but it sometimes fails completely, is sometimes damaging, and almost always takes a long time. Medication, by contrast, has a more guaranteed effect in a short time, but acts globally (contrary to popular opinion), and can have serious side-effects in the short and long term. (In fact one of the most important drives in pharmaceutical medicine at present is improving targetting, that is making drugs that will home in on specific areas rather than being distributed generally.) Let me put this another way. Suppose that I am greatly distressed about something, unable to sleep, and generally failing to cope. A major tranquiliser or sleeping-pill may provide instant relief by acting in a blanket fashion on my brain. It will make me less anxious, but it will also impair my reactions, make me less competent to drive, and lower my whole performance. (These side-effects are not generally recognised.) There are two reasons for this lack of specific action: we are unable to target drugs accurately; and we do not understand which areas of the brain to target them on. "Talking it over" acts quite differently: although we have no idea how it works, we simply rely on the usual performance of the brain to target it properly. Experienced counsellors can pinpoint an area needing treatment, and with no knowledge of neurophysiology they can address whatever it is in the brain that is disturbed by using everyday speech. The everyday processes which convert sensory stimuli into feelings and observations are harnessed to reform those feelings in a less distressing way. Often the areas requiring treatment are well defended, particularly by rationalisation and denial, but once broken down the change for the better can be spectacular, and can involve a huge outpouring of repressed feeling taking the form of tears or laughter.

It is therefore an illusion to suppose that we can always express in words all the nuances of our dispositions to argue for and believe certain things or to perform certain actions. We often call such unaccountable beliefs or actions "intuitive", because although held very deeply they defy articulation even in our own innermost thoughts. Oddly we have few misgivings about crediting a doctor or engineer with skills arising from their dedication to a life of study and practice (feedforward), but many such misgivings where we are asked to take seriously the intuitions of others who in their less formalised ways have also spent their lives consciously and unconsciously selecting the environments in which they would live and by which they would be shaped. It is only a spurious objectivism which regards the first as legitimate and rejects the second on the grounds that the first is examinable expertise and the other unquantifiable. In fact it is this objectivism which conspires, wittingly or otherwise, to deny the personal nature of knowledge and commitment by undermining the credentials of experience as it generates the wisdom which passes beyond anything expressible in words. The replacement of the wise old woman with the general practitioner is not an unqualified gain.

REASON AND TRUTH

It may be taken as a symbol of the disjunction of the rational and the emotional that truth is largely regarded as propositional in our science-dominated culture, and as such detachable from life as surely and easily as we can detach ourselves from our lies.

The propositional notion of truth owes its dominance to what Richard Rorty has called the transcendental pretence originating in a movement of philosophical thought which culminated in Kant. This led, by its insistence upon the impersonal absoluteness of truth and the suspension of belief, to disparagement of the particular world-orientations which inform other cultures. But Rorty, unlike Polanyi, does not see truth as personal, as arising from relational aspects of beings who are alive; instead he sees it as conversational, as arising within the universe of commensurable and incommensurable discourse, and as such fundamentally tied to language in all its forms (cf. *Philosophy and the Mirror of Nature*, especially Part III). Polanyi by contrast understands truth as it arises in the uniqueness of our selves in faithful relationship with the other, whether the other person, the other which is the universe we seek in our enquiry to understand, or the other which we call God.

Reason has always been concerned with discernment of the truth and the separation of truth from falsehood. Such truths embrace truths of statements, truths of action, truths of principle and truths of being. When our understanding of knowledge and truth is transformed from the impersonal to the personal, transferred from the realm of the intellect conceived as separate from the realm of the physical and practical into the realm of living enquiry conceived as an integrated orientation of all being, reason too must undergo transformation and translation. It becomes inseparable from the interactions of persons in their living, and so the criteria by which we decide what is or is not reasonable must also be personalised.

It is vital to understand the precise distinction here being drawn between personal and subjective reason, which hinges upon the notion of responsibility to the other as exercised within a critical peer-group shaped by and orientated towards a common concern. We act and believe and reason neither as we please (arbitrarily) nor as some external power insists that we must (slavishly), but as we feel that we must in our relationship of personal responsibility to the object of our reason. The combination is typically one of personal conviction and public witness (that we are called upon to search for the truth and state our findings). An opinion is subjective when I make no attempt to eliminate irrelevant bias from it, adopting it out of arbitrary preference or prejudice and justifying it with self-centred reason. "Subjective" does not therefore mean simply "mine" or "originating from an individual". An opinion is objective when I pretend that I have eliminated all personal coefficients from it so that there remains not the slightest element of my own judgement in it. "Objective" does not mean simply that an opinion is universally binding or agreed upon. An opinion is personal when it arises from a steadfast attempt to eliminate unwarranted bias and, whilst originating in an individual, is held to be binding for all, that is held with universal intent. A subjective opinion is an opinion held, in the last resort, for my own benefit and in response to my own arbitrary whim; a personal opinion is an opinion held only within the bonds of faithfulness to the other, and as such held in the last analysis not for my own sake, but for your sake and for the world.

The personal nature of Christianity does not therefore arise from an act of defiance: "I will believe regardless of all arguments to the contrary"; it arises in an act of responsibility: "I can do nothing else; I have seen with my eyes and heard with my ears, and cannot wash the sight or the sound away". This conviction and the associated sense of obligation outweigh all the inadequacies of the arguments that can be put forward in its defence.

CONFLICT

The theological imperative that we take the other more seriously than ourselves invests us with responsibility to God and for our neighbour. To perceive is to become responsible (cf. Ezekiel 33:1-6). The more effectively the Other is proclaimed, the greater will be the conflict between Christianity and the world.

It is interesting in this context to examine some of the mechanisms by which conflict arises and is resolved. If strongly-held positions are fed as much by emotion as fact, occasions on which real discussions capable of altering the minds of the participants occur are very rare. The emotional investment of the parties in their positions will be strong so that they are immunised against arguments presented by the other side. Situations will become charged with such emotion that even if one of the protagonists did suddenly "see the light" it would be felt to be virtually impossible to "climb down" (a phrase significant in itself) for fear of loss of face. In other words, we each recognise that our society sees it as a disgrace to be found to be (or to admit to being) in the wrong. We prescribe "cooling off periods" during which we hope that some of the respective arguments will be given a chance to penetrate the layers of prejudice and defence that initially prevented them from being heard. Nobody enjoys being refuted in an argument, still less having to admit to it. Our reactions are often of anger and resentment, for whilst we may recognise that truth has been the victor (that we have been shown to be mistaken), we are nevertheless encumbered by a self-image of such fragility that to admit to our error is unthinkable. In common but extreme cases we may even go away and invent further spurious arguments to rebuild our shattered arguments and defences.

The fault is not entirely with the loser in this conflict. It is also noticeable how much most of us enjoy demolishing the arguments of others, and doing so in ways which virtually guarantee that the response will be the one described. The more vindictive the refutation, the greater we know will be the consternation of the vanquished, and presumably the greater will be our self-satisfaction. (I was once told by a mathematics don that the most vicious arguments at conferences are always between rival professors of logic!) The problem is as much how to be a magnanimous victor as to be a gracious loser.

This illustrates how fatuous are the claims that dispassionate reason must be the sole arbiter of truth and falsehood, for the conclusion cannot be drawn (as might be expected) that this shameful state of affairs serves merely to confirm that emotion should be kept out of argument and debate; it serves to show that emotion *cannot* be kept out of argument and debate, and how disastrous it is that we have failed as a culture to integrate our emotion and reason. (It is a curious confirmation of our inability to cope with victory and defeat in this respect that we preserve a clear distinction between "argument" and "debate": the former is a full-blooded, no-holds-barred, life-and-death affair with quarter neither asked for nor given; the latter is by contrast a light-hearted entertainment in which nothing is ever really ventured and nothing gained. The polarity of the ultra-serious and the ultra-frivolous is significant.)

There is therefore always emotional involvement in any argument, especially when cold logic operates, for that logic is intended to disengage the opponent from the emotion which is his source of energy. Emotion only clouds issues where it runs out of control; while under control it supplies a precise and accurate clue to the truth because it allows us access to the intuitions which are inaccessible to dispassionate reason. Emotion is said to generate conflict because it is associated with the "flight — fright — fight" pattern of response to threatening behaviour, but pure reason leads us to paint ourselves into corners from which only an emotional outburst will free us. As Asimov writes in the *Foundation* trilogy, "violence is the last refuge of the incompetent".

Our culture teaches us, formally and informally, that emotion is bad and reason good. Consequently we devote most of our lives to the education of our reason, and ignore the education of our feelings and emotions. We take them either to be inherited or to be unalterable, and certainly not to be educable. Is it therefore surprising if our real, powerful, but untutored emotions burst out inappropriately and out of control? But it is quite unfair to blame or denigrate emotion for that failing; the failing lies in our excessive expectations and isolation of reason, in other words in our neglect of the *personal*.

In a saner world we would rejoice whenever we were shown to be mistaken, thanking those responsible for removing the cataracts from our eyes, and we would be equally thankful as the victor that, so to speak, light had dawned in another soul (although the terms "victor" and "vanquished" would cease to be appropriate). If we regarded mistakenness as an inevitable part of learning, rather as the carpenter sheds no tears over the shavings on the floor, we would be better able to see that we often learn more from our mistakes and from relinquishing our most dearly-loved beliefs, for which we have fought hard and long, than from all the truisms we never question.

This partly explains objections to evangelical crusades: the sight of many hundreds of people going forward in a public admission of conversion, of having a change of mind, is frankly rather unpleasant, for it resonates with our own deep anxieties about admitting that we have in the past been mistaken. On the other hand, for many Christians it would be hard to cease to believe if it involved a public admission of mistakenness. It is easier to "go along", neither hot nor cold. It is easier to be "civilised" about religion and keep quiet about it than to be firmly for or against it, which is regarded as rather bad form. Odd as it may sound, real belief must face full-square the possibility of its own error; it must be prepared to enter into the realm of unbelief. If it cannot do this; if, as we so often hear, it must cling to itself for fear lest it should slip away, then it is not faith centred in the other but faith centred in the self, and as such unfaithfulness. The hardest thing about being a Christian is that there is nothing whatever to be done about it. A faith which we hold on to in our own strength is no faith at all. "But surely", will come the reply, "if there is nothing to be done then people are not responsible for their belief and there is no merit in believing?" To which the answer is simply, "exactly so". But the absence of merit is no difficulty once we appreciate that neither is there a reward, except that of knowing the other from a centre within itself, and as such knowing oneself in truth, in participation, in fullness of life.

PURE AND PRACTICAL REASON

We saw in our discussion of the Method of Doubt that its strict application leads to an uncomfortable division between the assumptions we need to make in our everyday lives, and scepticism. It is therefore obvious why a distinction between pure and practical reason developed. Mathematical precision involves us in an unreal world whose unreality is obscured by the limitations of our perceptions; methodical doubt involves an unreal world in which we choose to doubt many things life forces us to believe. Pure reason places us in a position where we apparently have to choose between certainty and reality. Einstein, borrowing an insight from Clerk Maxwell, wrote "as far as the propositions of mathematics refer to reality, they are not certain; and as far as they are certain, they do not refer to reality" (in "Geometry and Experience"). But reality always seems to hold the "whip hand", for it is the real world that we must live in, despite the fact that ever since Greek times mathematical symmetries and certainties have seemed more perfect than the world around us.

It offends some of us that the world is not more mathematical even than it is, and we are tempted to regard the world as imperfect as a result because our criteria of perfection are inspired by pure reason. We find ourselves drawn to the despairing conclusion which Vico put so succinctly, that the only things we can ever know perfectly are the things which we have *made*, as contrasted with those things which lie beyond such knowledge, namely the things which we have *found*. Because the things we make are more under our control than the things we find it is always tempting to conceive of the world in terms of our own artefacts as we have noted with respect to measurement and "cybernetisation". The dreadful danger is that the systems of control will become so powerful that breaking out of them again will prove impossible. We will be locked into the values we build into our own machines and theories. Rorty describes this nightmare in terms of the end of incommensurable discourse, the coming of a time when everything that can be said has been said and there is no new thing under the sun.

P. F. Strawson suggests in his *Logical Theory* that formal systems must be supplemented to account for everyday discourse. Despite not being as systematic as formal logic, such a departure from traditional practice would challenge our intellectual powers along lines suggested in Clerk Maxwell's conviction that mathematics must be re-embodied if it is to move closer to an adequate capacity to model reality. We have no idea what such a mathematics would be like, but we are becoming increasingly aware of the deficiencies of abstracted mathematics, which is always susceptible to the fallacy of misplaced concreteness.

Logic deals with necessary truths, as Wittgenstein observes in the *Tractatus*. But everyday language makes use of concepts which lack definite boundaries, and for which no sufficient conditions can generally be found. Under what conditions does an object fall under the concept "... is a cat"? What are the sufficient conditions for cat-ness? The open texture of language means that we can never have comprehensive knowledge of the referents of a sign (i.e. an object, in Frege's terms). Logic allows proof and contradictions because it is limited to one language stratum, but (as Waismann has pointed out) between different levels (e.g. between laws and observations) logical connectives are inappropriate. An experiment never offers logical proof or disproof of an hypothesis (cf. the remarks on Hume above); it merely adds to or detracts from the probability or plausibility of the hypothesis as assessed in the end by indefinable qualities governed by practical reason. Emphasising correctness is symptomatic, thinks Waismann, of those who have nothing to say. But we should heed Pascal's warning:

> Mathematicians who are merely mathematicians therefore reason soundly as long as everything is explained to them by definitions and principles, otherwise they are unsound and intolerable, because they reason soundly only from clearly defined principles.
>
> And intuitive minds which are merely intuitive lack the patience to go right into the first principles of speculative and imaginative

matters which they have never seen in practice and are quite outside ordinary experience.
Pensées, Penguin edition, p. 212.

In other words, knowing and being involves not just a double focus upon analysis and insight, but an interpenetrating dialogue which involves the deepest levels of the person with the deepest structures of the world of things and ideas which he explores.

PARTICIPATIVE REASON

Participative reason has often been associated with attitudes which are other-worldly and dualist, particularly in aspects of Greek thought, gnosticism, and Christian spirituality. But our emphasis upon the other as inclusive of the world and neighbour as well as God indicates that something very different is being envisaged here. In contrast to a piety which retreats from the world into contemplation of the divine and which acts in witness only insofar as it encourages others to follow its example, participative reason leads to our being thrown with renewed interest and energy into the world in such a way that the fundamental orientation of our interest is changed. Rather than being concerned with the world for our own sakes, concupiscently, we become fascinated by it for its own sake; it becomes a medium of inexhaustible richness. Thus, in contrast to the ancient gnostic idea that salvation comes by acquisition of knowledge that would serve the soul in the after-life, and the neo-gnosticism that prizes objective knowledge for similar, if secular reasons, participative reason rejoices in its communion with the real world for no ulterior motive, but out of a deep appreciation of all its wonder and diversity. A transformation is envisaged which shows, in the sheer difficulty we have in conceiving of it, how far we have travelled along the road to despair; it is a transformation of value in which reality is sanctified and enabled again to proclaim the goodness of God through its own goodness. Our eyes are opened to the pettiness and irrelevance of the selfish ambitions that lead us to sell ourselves in the quest for a fullness of life that will always evade us. Suddenly we can see the lies that we tell and the deceptions that we practise as themselves the sources of the false realities that oppress us, and we begin to understand that the ladders we climb in our own strength and for our own profit have no tops and no endings; they lead only into the dark.

To be reasonable is to participate in a process that involves far more than oneself. To be fully reasonable is to participate in the wisdom that governs the universe, and to perceive oneself and the world from a vantage-point which clarifies the contradictions and absurdities that perplex from lower levels. It is to achieve a focus which is both corporate and individual in that it values and cares for the individual in community and nurtures community through individuals. It is to break down the prejudices which have led us to conceive of ourselves in individualistic terms, and to be virtually incapable of conceiving of ourselves in any other way.

Our individualistic outlook can lead us to interpret the death of Jesus as the supreme self-sacrifice of an individual for the cause of truth. Jesus is often held up as an example to us on such a basis. We think of his last hours in terms of loneliness and isolation, seeing the forces arrayed against him as vastly superior in number and strength. And no doubt as an individual contemplating pain and humiliation and death he did act as, and have all the fears and failings of, an individual. But to leave the matter there would be to miss a more important aspect of his death, that he died the single death of an individual man out of a full sense of his participation in a greater process than can be embodied in any individual life. The Alpha and Omega, the Creator and Redeemer, bound up as he was with a movement of the divine will spanning all of created time, was at once a part of and the centre of all things, and as such greater than all the forces massed against him. The mystery of the incarnation which the Fathers of the Church laboured so long and mightily over must be understood in terms of the unity of divine and human action which made possible the part Jesus played as man in the redemptive activity of God. This man, by virtue of his faithfulness to the divine will, to the processes shaping the universe and bound up with the divine purpose from the outset, was able to be at once in the midst of trouble and pain and at one with the greater cause for which he fought and died.

It ought to be possible to speak of this orientation of the person to the process of divine creativity in terms of prayer, but the word has become so eroded by being excessively burdened with connotations of self-centred intercession and other-worldly meditation that it is almost impossible to use it. Prayer as the means to strengthen and direct our lives in the real world, as a complete attitude toward the purposes of God, is so unfamiliar to our selfish and utilitarian world that it might seem that it could never be rejuvenated. Nevertheless the need is clear for just such a rediscovery of prayer, for no-one can encounter the cynicism and scepticism of the world in his own strength and fail to be diverted from his intended path, especially when the world refuses to believe that there is any source of strength and direction to be had which does not originate in itself.

THOUGHT

For any kind of education to be successful in the long term the focus of the student must move from mechanism to understanding. Why is this conceptual revolution resented and resisted? Thinking is an effort, and the process of genuine thinking is resented because thought has the power to invoke and demand change, because what we do not understand convicts us of failure, and because our fear is that despite risking further effort we may still fail, thus making ourselves feel even greater failures. The cycle is self-perpetuating. Thought has the power to dismantle our existing patterns of understanding, our stable systems, by bringing us to the point where we enter into the realities and confusions of different world-views. To think is to risk mental disorientation akin to that which accompanies moving house: what was on the shelves in neat and familiar patterns is suddenly deposited in a heap in the middle of the floor. The stress arises from this disorganisation of the things by which we measure ourselves as if, to change the metaphor, a lighthouse shown clearly on a chart were suddenly to be moved. We realise that things are not as simple, straightforward or self-evident as they once appeared.

The social conventions which constrain religious and political discussions are based upon the recognition that such matters concern the protagonists deeply, that shifting ground is hard at the best of times and virtually impossible in these two areas, and therefore that discussion is likely to lead to ferocious disagreement. This ferocity reflects our feeling that to surrender these systems of ideas is to surrender our own security; we pretend often that we are defending our religion, our God, our party or our country, but more often than not we are merely defending ourselves. This is one of the reasons why the remarks made above about Christian faith not being something we cling to are of such importance. By insisting, as it has in all the best traditions, that faith is something that we need neither earn nor hold on to, the Church has pointed to the freedom thus generated for its members to enter fully into the worlds of others, confident that their own security does not depend upon their own tenacity but on the truths established and maintained by God. The archetype of this self-less confidence is the Son of God's own descent into the world of temptation and sin, pain and death, unbelief and corruption, in the incarnation, not thinking it necessary to cling to equality with God, but humbling himself, being made in the likeness of men, and becoming obedient even to death upon a cross.

The contrast between mechanism and understanding, between the formal and the unformalisable, between self-centred and other-centred reason, can be understood in terms of this defensiveness. If I enter into knowledge of a subject which remains mechanical (and thus objective) I do not risk being changed by what I learn; but if I enter fully into the subject in order to attain to a focal vision, in order to attain to understanding, I run the risk of being changed, of being required to think through again my former understanding and to share responsibility for changing myself. To enter into understanding is to be changed by what one comes to know. Herein lies one of the deceptions practised by objectivism, that in objective knowledge by remaining essentially detached from what is known mechanically and formally, we are engaged in a process of self-preservation, of immunisation against change, which prevents us from having to acknowledge that we have until now been wrong. We allow ourselves to pretend that we are not responsible for what we know, or for the kinds of people we have become as a result of the influences to which we have subjected ourselves. The greatest lie of this sort is the transcendental pretence by which truth itself is rendered static and absolute, and as such essentially detached and impersonal, knowable by dispassionate propositional reason of a kind that does not alter the knower. Contrast the Truth which is personal, which engages with us in our searching, and with which we form a relationship in which we know without ceasing to be responsible for what we know, and find ourselves transformed by its living, dynamic, creative power.

PLAYING AND EXPLORING

Of all subjects, mathematics probably has the greatest power to induce desperation, frustration and resentment. For those who "see", it is easy; for those who do not "see", it is impossible. In teaching mathematics one is always aware of the temptation to abandon the attempt to teach understanding, and to be satisfied instead with mechanism, with teaching by rote: in this situation do this and this and this. As one teaches one can almost see the uncertainty growing, the desperation gathering momentum as the processes involved in understanding demand that existing convictions are abandoned. In these states of disassembly pupils will complain that they "don't understand anything", and will begin to make elementary mistakes of a kind that hitherto they were never tempted to make. The process is disagreeable, for as with all change the fear is that there will never come a time of reassembly, of renewed and greater wholeness. And as is always the case, in the face of adversity we are tempted to flee back to safer and more familiar ground. For some teachers, of course, the undoubted fact that in some examinations rote learning is successful in terms of marks will be proof enough of its rightness. But if education looks further than this, to the kinds of skills that pupils will carry from school into later life, there can be no doubt that it is what we come to understand that endures, not what we con by rote to cast into the examiners' teeth. The problems of teaching understanding are compounded because courses work sequentially until shortly before examinations. What could be handled in "mind-sized bites", in Papert's wonderful phrase, suddenly becomes impossible when all the bites must be redigested simultaneously. Pupils will often complain particularly bitterly when examination questions are set which mix topics from different parts of the syllabus, since this offends against the compartmental learning that has preceded the examination. The ploy is, of course, deliberate, for examiners recognise that it is one thing to memorise a text-book proof — that can be done mechanically — and quite another to draw upon a fund of knowledge integrated together by understanding in order to adapt to novel combinations of problems.

Teachers in all subjects, to be good teachers, must be capable of entering into the confusion in their pupils' minds and finding for them a way to reorganise their understanding. But few teachers can actually do this, perhaps because they cannot trust themselves to relinquish their grasp on the correct method for fear lest they should be unable to find their way back to it again. A poor teacher will only be able to respond to an uncomprehending question by restating what had prompted just that question; a good teacher will be able to rely upon a richness of understanding which will permit him to adopt a variety of metaphors and analogies (that is, different methods of approach), some of which may make more sense to the pupil. That capacity for exploration is not only a valuable teaching asset; it also demonstrates to the pupils by example that the way to solve unyielding problems is to try to approach them from new directions, in other words to play around with them until they yield some new clues. But it is just this kind of playing, i.e. this unstructured and hence (for some) threatening departure from formal, mechanical lines of argument that demands the self-confidence to be wrong. It is so easy to present pupils with standard, pre-digested solutions to problems which disguise the heart-searching and the blind alleys which were part of the solution, and yet we all do this because we recognise that in our culture to be seen to be in error is to lose face.

Borrowing once again from Polanyi we can describe this extraordinarily counter-productive practice as pedagogical perfectionism leading inexorably to pedagogical inversion. Mathematics offers perhaps the worst example of this process. What pupils need to learn is not how to proceed when solutions are known, but how to proceed when they elude us, how in other words to do what Polya calls *heuristic*, how to perform the "finding out" process. Yet it is just this (for admittedly very good reasons) that teachers are reluctant to display. The perfection of the solutions proferred leads to the inversion of the learning process.

We see therefore that clinging to the coherence and self-sufficiency of a system, in other words to its internal perfection, does not enable us to acquire the essential skills we need if we are to learn to cope with change and unfamiliar territory. We recognise that children need to be supplied with "safe" areas to play in, not simply to protect them from coming to harm, but also because their playing must not be inhibited by unnecessary dangers which will discourage them from setting out into the unknown. This contrasts sharply with the hostility of many learning-environments, in which just this same exploring is either castigated by impatient teachers or pounced upon by vindictive fellow-pupils. While it is undoubtedly necessary for adolescents to learn to survive this experience, it can also leave (and I suspect usually does leave) lasting scars which inhibit play and exploration for the rest of life. (The ways and means of the transition from playing to exploring have themselves been explored by Robin Hodgkin in his book *Playing and Exploring*.) What needs to happen, in contrast to the destruction of the confidence needed if we are to sustain our play, is that as we grow older, far from relying upon others to supply and protect the space in which we can safely play, we must gradually take over that responsibility ourselves by generating a world-orientation sufficiently resilient to ridicule to ensure that we are never afraid to be wrong (and, more to the point, to be seen to be wrong and to acknowledge that we are wrong), without making it so rigid that it prevents us from seeing when we are wrong. We need, so-to-speak, to internalise our play space so that we can explore external space.

For adults, coming to terms with being wrong is a process which mirrors bereavement. At first we hold to some opinion, we so to speak possess it. Then something happens to prevent us continuing to hold it; it is taken from us, and we experience loss, a loss as much of self as of the opinion possessed. Our immediate reaction is often denial: I am not wrong; there must be some mistake; I will rework the argument to show you where you have misunderstood (corresponding in bereavement to: he is not dead; you are speaking of somebody else; sooner or later he will turn up). Denial is followed in most cases by anger: who do you think you are to refute my argument (corresponding to the anger often displayed towards the innocent bringer of bad news, or the unfortunate doctor who is supposed to have a cure for death)? Anger, after a shorter or longer time, may then turn into depression as the awful realisation that the idea (person) is really dead grows upon us and we have to come to terms with life without it (him). (Notice that at this stage there may be nothing substantial to replace the lost or refuted opinion or person, so that the typical state is "I can't go on!"; the future, which we had mapped out in our minds with reference to our opinion or person as a landmark, has been shattered.) Then gradually depression gives way to acceptance and reconciliation as new possibilities emerge, the wound heals, and we begin to contemplate a new future with new maps gradually forming in our minds.

Our discussion of the nature of reason has led us into some unexpected territory. We have come to see the importance of locating the centre of our lives outside ourselves, and reason as one of the means whereby we emerge from a centre in ourselves to be changed as we come to participate in the other. We have seen how reluctantly we relinquish our selfishness, and some of the reasons and remedies for this reluctance. And we have located one of the central themes of our cultural pathology in the unacceptability of being seen to have been wrong. In all this we have presupposed that there is some other centre and that to know it and participate in it is desirable for our individual selves as we become persons, and for our societies as they become communities. The question of the nature of that other centre must now become our focal concern as we move from considering the nature of reason to exploring the nature of truth.

REALISING TRUTH

IT is easy to speak of truth in terms of *knowledge*, then to conceive of knowledge in terms of formal statements, and therefore to equate truth with truths of statements. But why do we tend to speak of *knowing* the truth rather than *feeling, believing, sensing, understanding* or *being* the truth? Such expressions seem strange in varying degrees because of a cultural bias which tends to deny that the inexpressible can be true, and that bias arises from a conviction that anything true can be put into words. A vicious circle ensues.

It is easy to ask questions as though all questions are good questions, then to find that the object of our questioning is incapable of providing a satisfactory answer, and to conclude that it is somehow in the wrong, rather than the question itself. We therefore tend to restrict the kinds of questions we ask of ourselves and the systems in which we believe, while allowing unbridled scope to the questions we ask of those systems and people with which we disagree.

It is easy to ask questions under the impression that we would recognise an answer if we encountered one; but an answer would need to be true in order to qualify as an answer at all. Therefore if our question is "what is truth?" it seems that the criteria by which we would assess an answer are themselves uncertain.

The assumption that we would always recognise an answer to a question when we saw one can be shown to be false: most important answers demand a shift in our understanding which can only come about after prolonged exposure to their implications or because those answers provide us with an integrative insight which transforms our understanding instantly. Generally speaking, however, the thing which commends itself to us about a supposed answer is more likely to be its promise than its self-evident truth (in Kuhn's terms we become open to the possibility of a change of paradigm). A simple illustration from mathematics will suffice. It is a common experience in mathematics to read a proof in a well-respected text-book, authenticated by the recommendation of the community of mathematicians, and yet not to see that what it claims to prove has in fact been proved: what clearly must be an answer is not perceived to be an answer because some of the conceptual apparatus needed fully to appreciate it is defective.

These three brief observations may be summarised in the three questions: what is truth; what is a true question; what is a true answer? They can be phrased differently: do we know what truth is; do we know how to enquire after it; do we know what an answer to our enquiry will look like?

In mathematics we are far from a clear understanding of the nature of truth unless we restrict it to internal truth, i.e. consistency, coherence and theoremhood; but Gödel has shown that we cannot prove that rich formal systems are consistent. We know how to check proofs mechanically, but we have no idea how to decide whether a particular well-formed statement is a theorem, or how to generate a proof for it. From different sets of axioms we can generate different systems of theorems each of which, by criteria restricted to internal consistency, is as true as any other; but we have no idea how to decide which of them is true in any wider, external sense, and some have ducked the issue by denying that there is any such external sense of "true".

Why do we regard the question of truth as of such importance? Some people do not: usefulness has replaced truth as a criterion of value. We have seen that mathematics must make assumptions and approximations if it is to provide us with models of the world which we can use and solve, and that these models undoubtedly introduce distortions. But the fact that we cannot handle greater accuracy, and that these approximations work, makes these limitations of smaller significance. In discussing measurement we acknowledged the loss of information involved in reducing reality to number, and in the problem about Tom's age we deliberately destroyed or ignored detail in order to reduce the problem to a form amenable to mechanical solution.

> Descartes algebrized [sic] geometry. Algebra is specifically a matter of getting rid of some content. Hence in virtue of Descartes's discovery, geometrical proof can be conceived as purely formal.
> Hacking, "Leibniz and Descartes: Proof and Eternal Truths".

The successes of the approximations science has made cannot be denied, but are those successes sufficient to silence the demand for truth forever, and should they be? A clue to the answer lies in the observations we made above about the eventual need to look carefully at our foundations, a need which mathematicians have felt strongly over the last century, and which scientists are only now beginning to recognise as some of the implications of completely unfettered scientific research are becoming clearer. Over the last four centuries a problem has arisen. It is not the threat of nuclear holocaust, or genetic engineering, or pollution, serious as these all are; it is more fundamental than these: it is the question of the self-understanding of mankind in terms of our place and significance in the universe. In particular, the question arises which is *true*, a scientific description of the human situation in terms of a chance irruption of life on a minor planet of an average sun at one end of an average galaxy, or a religious description of the human situation in terms of the culmination of divine purpose in the creation and nurture of children able to become God's sons and daughters. The battle-lines are drawn between a *quantitative* and a *qualitative* description of human significance.

An immediate dilemma emerges: it is one thing to argue that we *need* to believe in our ultimate significance in the universe, and therefore that we need to believe in a divine Creator whose purpose we fulfil (despite the fact that science is really right all along); it is quite another to argue that it is *true* that we are the fulfilment of the purpose of a divine Creator, regardless of our perceived need for such a belief about ourselves (and therefore that whatever science says about us is only part of the truth). Or, to put it another way, there are plenty of people who argue that things are getting worse so we need a revival of religious values (regardless of their truth or validity), which is a form of pragmatism or utilitarianism; there are far fewer who demand a revival of religious values because they are *true*, and as such to be taken seriously.

In saying all this I have deliberately adopted terminology consistent with the belief that truth is something ultimate and binding, that in the face of truth all else is condemned as error and falsehood, and that human life depends upon acknowledging and obeying that truth. I have done so in order to start from the top and come down *because it is logically impossible to start from the bottom and build up*. In other words, it would be wonderful (I speak as a fool) if we could begin from some uncontroversial premises, do some logic, and arrive at the conclusion "therefore truth is something ultimate and binding", but it is also impossible. In fact, as I have already argued, it is intrinsic to the nature of the theological imperative that this be the case.

Odd as it may seem, just such a view of truth is held by many mathematicians, and can be seen to underlie the practice of mathematics (as well, of course, as other disciplines). However formalistically we behave and explain our behaviour (e.g. "I just push formulae around"), we study mathematics because we perceive the intrinsic worth of the systems under study (I discuss one exception to this below). We may only regard them as true in themselves in a rather limited way, but our conviction of that self-justifying truth is nonetheless real.

TRUTH IN MATHEMATICS

This section is somewhat technical, but I have attempted to steer a course between excessively mathematical concepts and triteness in order to give the reader the flavour of the subject and its importance. Those interested in further reading should consult, for example, J.N.Crossley's excellent little paperback in the OUP Opus series called *What is Mathematical Logic?*

We have already seen that mathematicians late in the nineteenth century began to look at the foundations of mathematics in order to set their house in order before confusion resulted from examination of alternative logics and systems of axioms. The "big names" are Boole, Frege, Russell, Hilbert, and Gödel. What did they discover?

Mathematical systems come in various shapes and sizes. In general, as the complexity and power grows so does the difficulty of establishing fundamental theorems; consequently questions about the nature of truth can be answered in simple systems, while remaining unanswered in more complex ones. Moreover, the most general systems of mathematical logic (logical calculi) pay a price in abstraction for their generality. For example,

If A implies B, and B implies C, then A implies C

Although this looks true, and seems as if it would remain true whatever A, B, and C were replaced with, it actually *says nothing*, i.e. it is devoid of content. Somehow we have to decide *whether* A implies B, and *whether* B implies C; thereafter the theorem enables us to infer that A implies C.

The question then arises whether there are any limitations to be placed on the kinds of things A, B and C can be replaced with? Crossley puts the matter thus:

> [But] when we just have a collection of symbols, they are susceptible to many different interpretations (or models as they are called) and it is possible that interpretations exist that are entirely different from the domain we thought of in the first place.
> *What is Mathematical Logic?* p. 6.

Löwenheim and Skolem proved that there were such interpretations in about 1915. Later Gödel showed that the predicate calculus is *complete*, that is that every well-formed statement in it is a theorem provable from the axioms. This is exactly what we want from mathematics, so everything seems fine. But the predicate calculus is extremely general, which means that it is intentionally true in every interpretation at the price of being devoid of content (which must be supplied by the application). It is so general, in fact, that it cannot even be used to do arithmetic (since arithmetic is not devoid of content).

Once we step up the power of the system so that it can deal with arithmetic two things happen: first we have to abandon the notion of complete generality by restricting the domain of validity of the system (in other words we have to be careful what we replace A, B, and C with in the new theorems); second the completeness result breaks down. Not only does it break down: its converse can be proved, that there are well-formed true statements in the theory which are not provable. This is the celebrated *Gödel Incompleteness Theorem*, proved in 1931. Gödel also showed that you cannot prove these richer formal systems to be consistent either from within those systems.

A great deal has been inferred from these results (a lot of it unjustifiable and extravagant), but they do certainly point to one important truth: there is a necessary trade-off in mathematics between the absolutely general, complete, closed and provable (on the one hand), and the more specific, incomplete, open and unprovable (on the other). Moving in the direction of open-textured formal systems and away from precise symbolism introduces an inevitable loss of provability. Nagel and Newman, commenting on a result known in the nineteenth century, express the matter thus in their book *Gödel's Proof*:

> [It] called attention in a most impressive way to the fact that a *proof* can be given of the *impossibility of proving* certain propositions within a given system.
>
> op. cit. p. 10.

The result in question was that Euclid's axiom of parallels (that through any point outside a line one and only one line can be drawn parallel to the given line) cannot be derived from his other axioms. (Riemann replaced this axiom with another asserting that through such a point no line parallel to the given line could be drawn, and in so doing provided the mathematical basis for Einstein's General Theory of Relativity.)

TRUTH IN LANGUAGE

In the chapter on reason we saw how concepts are not like photographic images (mirrors of nature, in Rorty's phrase), carrying the truth within themselves, but indicators and communicators of meaning which must be interpreted creatively by the observer or recipient (just as they are generated creatively by their author). Polanyi's concept of indwelling provided the key to this process, coupled with his distinction between subsidiary and focal awareness. Many philosophers write as if truth is reducible to the truths of sentences, as if knowing all true sentences would involve knowing all truth. Now it is probably true that knowing the truth involves being able to recognise true sentences, but it is certainly false that knowing all true sentences would afford comprehension of all truth, since that would eliminate the ineffable from truth. This could only be the case if we deliberately restricted truth to a propositional form. This description of indwelling affords a way of conceiving how we find our mind attuned to the trans-formal truths which demand our allegiance and invest us with responsibility; it enables us to acknowledge the possibility that we might come to encounter God.

We are under pressure to reduce the scope of truth to mathematical, rational, sentential, or empirical realms. Yet when we use words to communicate we do not attend to each word in a sentence, analysing the grammar, case-structure and tenses as we are encouraged to do in learning a second language (for a reason which is not entirely clear to me); we dwell in those words, allowing them to wash over us, while concentrating on perceiving their focal meaning. Similarly, we assess the accuracy and appropriateness of the use of language by a speaker or writer by how well he manages to convey his meaning to us rather than by analysing his syntax. By contrast, someone who is almost fluent in English, but whose usage is slightly unusual or incorrect, is often difficult to listen to or understand because we find our attention being diverted from the focal to the mistakes made at the subsidiary level. Something similar occurs when a musical performance is marred by wrong notes: we find ourselves making a conscious effort not to be distracted, not to lose focus. At its most powerful indwelling makes us completely oblivious not only to the subsidiaries which lead us to their focal meaning, but to our surroundings as well. We speak of someone being "gripped by" or "immersed in" a book when they are so involved with the story that they do not hear us speak to them. The marks on the paper and the words which make up the sentences become transparent as we enter into the world they create in our minds.

The question which arises from this is: where does the truth reside? Does it reside in the sentence (or formula) itself? In the mind of the author? In the mind of the recipient? Somewhere else? Clearly we are not seeking a spatial answer to this question; we are concerned to know what to examine (the sentence, our own mind, the author's mind, or something else) in order to be guided toward the truth of the sentence. The problem is not as easily dismissed as we might imagine, and again mathematics affords an illustration. It is easy to say that Australia was the largest island in the world before Australia was discovered (see above) because of our prejudice about physical objects. Hence truths about Australia seem to have a home. But mathematical concepts such as number and shape do not inhere in any particular object (in fact, they do not inhere in any object, as we have seen). Are they *found* or *made*? Either answer plunges us into a dilemma. A hard-nosed Platonism might lead us to the following account:

> Mathematical objects are independent of our minds and, unlike physical objects, do not interact with our bodies to cause alterations in our brains that lead ultimately to knowledge of them. But they must be postulated to account for the existence and growth of mathematical knowledge and, to the extent to which other knowledge is dependent on mathematical knowledge, of other knowledge as well.
>
> From Benacerraf and Putnam's Introduction to *Philosophy of Mathematics: Selected Readings*, p. 30.

They go on to say how repugnant such non-physical powers would seem to most philosophers and psychologists, but any solution which avoids them merely replaces one set of problems with another. If we reject the idea that mathematical objects are *found*, then we seem bound to accept that they are *made*. Then we have to explain (a) how an uncountable set can exist in the human mind, which is finite (the real numbers are uncountable); (b) how we are to choose between rival mathematical theories; (c) why mathematical theories have been so successful in empirical science. As these authors point out, the fascination of the subject is that nobody has any idea how to resolve such problems in an entirely satisfactory way.

TWO KINDS OF TRUTH

In his *De Veritate*, Anselm suggests that we distinguish two kinds of truth, namely the truth which belongs to a sentence when it is properly constructed (we would call it syntactic or grammatical truth), and the truth which arises from a sentence when it signifies according to its purpose, that is when it conveys the meaning it is intended to convey (semantic truth). We may speak of these as truths internal and external to the language concerned respectively (in mathematics the parallel is with well-formed-formulae — syntactically correct statements — and theorems — statements provable within the theory). The instance of the imperfectly spoken sentence which still conveys meaning shows that syntactic truth is not an absolutely necessary condition for semantic truth:

we can attain to the focal meaning even when the subsidiaries are defective, but it is more difficult. On the other hand it is our perception of the meaning of the sentence which is the source of our appreciation of its importance. We value a portrait because of the insight it conveys, and we distinguish valuable portraits from hack studies by their capacity to convey this meaning. Nevertheless, it is still possible to imagine a case where despite having failed to grasp the meaning of a portrait or sentence adequately we sense enough of its meaning to recognise its value; the subsidiaries are to us as the trail to a bloodhound: we know they lead somewhere important, but we are not entirely sure where.

If we turn from the appreciation of a sentence to its expression we see the other side of this mutual dependence. To express myself clearly enough to make my meaning available to you I must have a certain command of an expressive medium or language; but to have anything to say in that language I must perceive significant truths. Moreover, because thoughts occur in representations, and cannot occur without a representational system, unless we have command over a formal system our capacity to have significant thoughts will be curtailed. We must add to the mutual dependence of syntactic and semantic truths for the *recipient* another set of dependencies for the *author*: the expressive medium must be shaped by a fund of meanings which it is the author's intention to communicate; the conceptual world of the author depends upon a medium capable of carrying representations (thoughts) arising from that conceptual world of sufficient richness to feed and extend it. Sometimes we experience the frustration of an insufficiently expressive medium when we feel that a thought is striving to come to expression but that we cannot find the words. (This is one reason why we value a fund of quotations to draw upon which we feel express important thoughts in an unusually apposite and powerful way.)

DUAL CONTROL

The inter-dependence of expressive medium and conceptual world we have described exemplifies the phenomenon which Polanyi calls *dual control*. As a scientist he was especially interested in biology and the processes involved in evolution, and he used the concept of dual control to describe the ways in which the laws of physics and chemistry place certain limitations upon the kinds of organism that can exist, whereas the kinds of organisms that do exist in their turn place constraints upon the particular patterns assumed by physical and chemical processes. With this notion Polanyi anticipated a good deal of modern systems theory, especially feedback concepts. A particularly powerful (and potentially dangerous) example of dual control can be seen in the new science of genetic engineering, in which a species generated by a particular genetic pattern and process (from below upwards) has found the skills to interfere with its own genes (from above downwards). This seems to some rather like sawing through the branch you are sitting on because it is impossible to foresee all the consequences of altering even a single gene.

In most cases instances of dual control are more innocent and commonplace. We have cited the upward control exercised on what we can express and understand by language, and the downward control over language we can exercise using the concepts which arise out of language. This terminology assumes, as Polanyi does, that reality is stratified in a hierarchy of levels which relate to one another according to the principles of dual control. Where more than two levels are involved the process might be called multiple control, as when (from below upwards) chemistry and physics influence biochemistry, biochemistry influences health, health influences individuals and individuals influence societies, or when (from above downwards) societies influence individuals, individuals influence their environment, and their environment influences biochemistry, chemistry and physics.

In adopting this hierarchical model Polanyi was assuming a position fundamentally opposed to reductionism, or "nothing-buttery" as Donald Mackay has called it, the view that everything, being composed of molecules, atoms, or sub-atomic particles (it is a matter of some dispute which level we take to be "fundamental", if any) is "nothing but" molecules, atoms, or sub-atomic particles (in other words, that love, beauty, meaning and truth — amongst other things — are illusions). This controversy, between those who like Polanyi regard such a philosophy as an inversion arising from perfectionism, as a form of nihilism, and its adherents, bears directly upon the questions "what is reality?" and "what is truth?", and it raises philosophical issues of the greatest difficulty and importance, having serious implications for our self-understanding as human beings and therefore for the future of the human race.

One of the strongest arguments in favour of reductionism arises from the apparent asymmetry of multiple control in the biosphere. Reductionists deny the downward control exercised by higher levels upon the lower, and they are drawn to this by the apparent impossibility of regarding the laws of physics as malleable in anything approaching the way that social and psychological laws are. They acknowledge that dopamine deficiency causes epilepsy, or acetyl-koaline deficiency senile dementia; they are unable to acknowledge that there are attitudes of mind which affect the biochemical behaviour of the body because they are unable to find any mechanism whereby this might occur. There is, however, extensive evidence from pharmacology, based upon comparisons between real drugs and placebos (that is, tablets which are administered but have no actual drug in them) that the belief that a drug has been given materially affects the rate of return to full health (the same correlation can be shown in instances of good doctor-patient relationship, although the evidence is controversial).

If human beings, along with all other things, are nothing but sub-atomic particles, then the supposition that there is more reason to preserve the life of a human being than the existence of a stone is an illusion which arises from the distorted perspective of our self-interest. If, in other words, the one true reality is, *mutatis mutandis* in this quantum-relativistic age, only "atoms and the void", then everything which is not described in terms of atoms and the void is a lie (including the process of declaring it to be a lie, since concepts of truth and falsehood, being human, are themselves worthless and illusory). In other words, the reductionist demonstrates as a direct consequence of his perfectionism (that everything must be reduced and analysed to the indivisible and irreducible minimum) that he and his speculations are worthless illusions, epiphenomena floating meaninglessly upon an ocean of emptiness. He has, thereby, proved Polanyi's point that perfectionism leads inevitably to nihilism and inversion.

Our experience indicates that this is an untrue and unacceptable conclusion. Unfortunately we now lack confidence in our feelings to such an extent that we are reticent about rejecting such an argument on such grounds. This lack of confidence is unfounded. It arises from the domination of analytic, self-centred modes of thought over integrative, other-centred modes, from the wounds inflicted by the honed scalpel of the sceptic upon the logical fragility of the believer. But this vulnerability of other-centred reason is inescapable because it self-consciously turns aside from total reliance upon the strictest canons of logic in order to apprehend truths which it perceives to be more important. The self-centred rationalist is like the creature in Aesop's fable who grasps an apple in a jam-jar and then cannot withdraw his hand through the neck; because he can neither remove his hand nor bring himself to relinquish the prize he has won he is condemned to remain there for ever. The self-centred rationalist must observe the wilderness he has created and think "yes, it is a barren wasteland, but it is bed-rock, and it is certain, and I feel safe with it"; and he feels safe with it because it requires of him no risk, and costs him nothing (except the soul he does not believe he has).

A scientist may say that we are nothing but dust, and I object that I have a vision of the nature of man which contradicts that assertion as regards the "nothing but". A psychologist then reduces my vision to the outcome of some deep insecurity induced by my up-bringing and therefore to my inability to face the reality of my insignificance. I object that I would be prepared to face that reality but for the fact that my vision denies it, and he accuses me of arrogance in that I claim to have a vision denied to him. I reply that he seems unduly pessimistic; perhaps his pessimism is induced by an unhappy childhood? And so it goes on. Nothing can be established by such an exchange, but it restores the balance a little.

THREE MORE KINDS OF TRUTH

The two kinds of truth we have distinguished are insufficient to embrace all possible truths. A sentence could be syntactically correct by conforming to the grammar of a language, semantically correct by correctly conveying the meaning intended by its author within a culture sharing that language, that is, true in both the senses we have delineated, and yet nevertheless be false. How can this be? The matter is clearest in the case of mathematics. If we consider a well-known formula such as Einstein's

$$E = mc^2$$

then it is clear that it is well-formed (it does not break any of the rules of mathematical formalism); and we know that it arises from the special theory of relativity because we can prove it from the axioms and principles of the theory (it is consistent with the rest of the theory); but is it true in the sense that it properly describes the relationship between Energy (E), mass (m) and the square of the velocity of light (c^2)?

A third kind of truth must be invoked, in fact the oldest and most obvious kind of truth, which goes back at least as far as Parmenides who pronounced a statement true if it describes a state of affairs as it is and not otherwise. In other words, a statement is true if it corresponds with reality and not otherwise. This is the Correspondence Theory of Truth.

The Correspondence Theory is notoriously problematic, despite being intuitively appealing and straightforward, because it is far from clear what it actually means to say that a statement "corresponds with reality". One problem is that it is difficult to see how a statement can correspond to something that is not a statement. Another is that it is impossible to say that a *person* is true on the correspondence principle since there is nothing with which such a person "corresponds". Do we, then, simply mean self-consistent and coherent by "true"? Clearly not, because someone could be consistently false. Personal truth seems to require the notion of *faithfulness* to the nature of God, world and self which cannot be described in terms of correspondence. A *statement* can therefore be said to *refer us faithfully* to the state of affairs in the world; by understanding its meaning we learn how things are. A statement is true if its meaning and the meaning of the reality to which it refers are the same; or, if we choose to retain the earlier terminology, if its meaning corresponds to the meaning of the reality to which it refers. For example, the statement "snow is white" means something to anyone who understands English. Similarly, the physical phenomenon we call snow means something to anybody acquainted with it (regardless of which language he speaks). If what the statement "snow is white" means corresponds with what the physical phenomenon we call snow means, then the statement is true and not otherwise.

The unhappiest feature of this solution is the association of meaning with physical phenomena. It seems unnatural to say that the physical phenomenon we call snow "means" something to us which enables us to say whether it is white or not, but that is just the point of the notion of concept developed by modern linguistic philosophy, namely that to have the concept which "snow" names (to understand the meaning of "snow") is to understand whether or not it is white. (This has nothing to do with mental images.) The key to this puzzle lies in the observation we made in the chapter on reason that our concepts are not mirrors of the world, but interpretations of the world; in the terms used earlier, there are no "raw facts". To recognise that snow falls under the concept "white" is to establish an unavoidable connection between snow and everything else recognised as falling under the concept "white". We would not enunciate all those connections; nor could we if we wished, but they are there. Our understanding of the meaning of "snow" and the meaning of snow is inescapably coloured by its falling under a concept shared with other words and objects. It is the richness of the associations poets command and their capacity for presenting familiar words in new and striking sentences and phrases that leads us to value their poetic abilities.

It may look at this stage as if we have made a simple and straightforward concept impossibly complicated, but the distinction we have drawn between truth considered as a property of a sentence, and truth considered as a relational property between the meaning of a sentence and the meaning of reality is crucial to our liberation of truth from formalism, and the establishment of a third, referential kind of truth.

Because he regards all true action as action according to the will of God, Anselm introduces a third sense of "true" in respect of things which are as they ought to be. By insisting that the ultimate ground of truth lies in God, the Supreme Truth (Anselm is influenced by the Platonic doctrine of the Good in this respect), he implies that language, man's creation, must refer to rather than determine truth, and thus that any discussion of the nature of truth must, in consistency, be open to a ground beyond itself. It is possible for a proposition to be both syntactically true and semantically accurate without this dimension of openness, but only by being incorporated in an act of participation in the Supreme Truth will it achieve the third kind of truth, for nothing is true unless it participates in truth.

Our interpretation of the world is concept-laden. There are no "raw facts". Therefore the way we understand the world is theory-dependent, and the comparisons we make between the truths associated with the meanings of sentences and the meanings we discern in the world will depend upon some kind of fundamental orientation to the world, some theoretical perspective. What we will regard as true therefore depends upon our ultimate convictions, not simply upon supposedly impersonal and objective facts. From an other-centred perspective we will insist upon a fourth level or category of truth over and above syntactic, semantic and referential truth; this fourth level we may call moral truth. Such truth not only involves being as it should be according to internal rules, or expressing as it should express according to the intention of its author, or referring as it should according to the nature of the world, but being and entailing as it should according to the purpose of the other conceived as part of a greater whole. The sense of "entailing" here intended is that such truth should have consequences which are themselves true according to the same criteria.

A sentence could conform to the rules of a language, convey the intended meaning, refer correctly to the state of affairs in reality, and be in harmony with the purposes to which it relates, and yet still not be quite *right*. An author will often correct a sentence many times searching for the right form of words, not because any of the first four kinds of truth are absent, but because the sentence itself does not quite capture his full meaning to his satisfaction. This aesthetic sense determines the final adequacy of his work, and is governed by his sense of the true nature of what he wishes to say. His sentence must embody that meaning. It must *be right*. This fifth kind of truth is the truth of being.

We therefore arrive at the following five-fold classification of truth *with respect to sentences*:

1 — *syntactic truth* is determined by whether or not a sentence conforms to the rules of a language;

2 — *semantic truth* is determined by whether or not a sentence signifies what it is intended to signify;
3 — *referential truth* is determined by whether the meaning of a sentence corresponds with the meaning of the reality referenced by that sentence;
4 — *moral truth* is determined by whether the meaning of the sentence is as it ought to be according to principles discerned through participation in the Supreme Truth;
5 — *the truth of being* is determined by the adequacy of the sentence in its own right, by whether it is intrinsically perfect.

The accounts given so far bears upon truths of *sentences*. Can they also be applied to other formalisations? Take action: syntactic truth becomes *competence*; semantic truth becomes *effective execution of the intention of the actor*; referential truth becomes *action in accordance with an objective reality*; moral truth becomes *action bearing upon other actions and purposes appropriately*; truth of being becomes *intrinsic perfection*.

All five kinds of *formal* truth can only arise from an unformalisable truth which is part of the person of the actor, or the nature of the organism. Assessment and realisation of formal truth therefore depend upon a two-fold creativity, as remarked before. In a musical recital the performer and his audience act creatively to generate five kinds of truth: syntactically the performer plays correct notes and rhythms, and the recipient hears the formalism correctly (as is not the case, for example, when a westerner hears Japanese music); semantically the performer plays what he intends to play, and the recipient hears it as he intends it (which may not be the case if he is remembering other performances); referentially the performer is faithful to the composer's intentions while augmenting them with his own genuine insights, and the recipient obtains a valid insight into a facet of the composer's vision; morally performer and composer must be inspired by a valid insight into the nature of things, and recipient must be moved to such insight; ontically the piece and this particular performance must be worthy of their place in Creation and deserving of respect in their own right, and the recipient must relate to the music faithfully and respect that which arises from and is other than himself.

In all five respects performer and hearer must acknowledge the primacy of the other if their formal acts are to be true. Their *being towards* the other must be utterly faithful, fired by a sense of responsibility for that which is greater than themselves and deserving of respect in its own right. A possessive attitude to the music would involve interpreting and hearing it as they pleased, without reference to the composer's vision. The performer does not try to fuse his own being with the inspired creation of the composer, but stamps his own being upon it, denying the creative insight of the other. True performative acts involve the three stages of original authorship, obedient (but not subservient) transmission, and faithful reception. All involve a fundamental willingness to recognise in the other something which (a) deserves respect; (b) is not mine to do with as I please; (c) has in its otherness the capacity to enrich me; (d) invests those who perceive it with responsibility for it. If the performer and hearer do not allow themselves to come under the "spell" of someone else's vision or something else's reality their eyes can never be opened or their own vision enriched by it. If they remain *outside*, that vision will never be theirs. They will be unable to transmit their perceptions to others. All they can do is to share the poverty of their individuality, because they have questioned the music, but never been questioned by it.

The person of the musician is shaped by the music within which he dwells. By feedforward systems, those who commit themselves to music will be shaped by music; those who commit themselves to science, mathematics, theology, history, will all be shaped by them. But there are *so many* things that we could choose, and our time is finite. How can we make, when young, the choices that will best enrich us and the world through us when we are older?

THREE KINDS OF THING

These suggestions lead to the following scheme. Let us suppose that beside our five kinds of truth we place a three-fold classification of "things" which are potentially true. These are:

1 — uncreated things   (God);
2 — created things   (found things);
3 — creations of created things   (made things).

Uncreated things are not contingent in that they do not depend for what they are on other things — the only candidates we have for such uncreated things are God (if we are theists) and nature (if we follow Spinoza). Created things enjoy a first-order contingency in being dependent upon uncreated things. Creations of created things enjoy a second-order contingency in being dependent upon created things. (If computers ever display something which could be regarded as genuine intelligence it will be necessary to add a fourth category, but we need not pursue that topic here.) The world falls into both the second and third categories, being partly found and partly made; our intellectual systems fall entirely within the third category, and therefore tell us about the second category only insofar as our understanding has the power to shape how we see reality truly, and to manifest itself in true formalisations.

These three categories provide the framework for an answer to the question "What is Truth?" in terms of an inclusive relationship between all three.

> Truth is that which is realised in a relationship of fundamental faithfulness to the uncreated, the created, and the creaturely creation.

The uncreated, not being contingent, cannot stand otherwise than in a relationship of faithfulness to itself. When we say "God is true" we mean that the uncreated stands in a relationship of fundamental faithfulness to his creation and his creatures. The created is true when and only when it is faithful to the uncreated, and manifests that faithfulness in its own creation. The creaturely creation is true only when it is faithful to the created which is in its turn also faithful to the uncreated. A bad tree cannot bear good fruit, or a good tree bad fruit.

FALSEHOOD

Two points need to be made about falsehood. The first concerns the extent to which falsehood is entailed in the concept of truth (i.e. whether there can be a concept of truth without a complementary concept of falsehood). The second concerns the upward contingency of truth in the five-fold hierarchy we have set out (e.g. the sense and extent to which a syntactically correct statement can be regarded as true despite its being a lie).

According to logic a proposition A always gives rise to another proposition not-A (the contradiction of A). By the *law of excluded middle* any syntactically well-formed proposition must be either true or false (that is, either A or not-A must be true). This law says, in effect that there is no third possibility, *tertium non datur*. But the possibility of falsehood only arises with the act of creation, i.e. the act of God in extending existence (which is his alone by essence) beyond himself, and in so doing creating the possibility that the creation might not be as it ought to be. Falsehood is therefore a feature of created things and the creations of created things (categories two and three above).

This raises a further set of questions concerning these two categories to which I propose only to draw attention.

> Concerning created things: what has theology to say about the concept of the *fallenness* of creation given that science discerns only cosmic order, and seems not to embrace any notion of imperfect nature; a virus or an earthquake may be a disaster from our point of view, but to science they are merely different phenomena to be explained; this is another way of saying that science is concerned with the *is* rather than the *ought* of the world; to science even the best and worst people (judged by human standards) are merely phenomena to be classified and explained.
>
> Concerning the creations of created things: by which criteria are

we to assess what we say, make and do: by the valueless criteria of phenomenology which describe without evaluating, classify without generalising, and observe without inferring, i.e. which seek to be completely value-free; or by the value-laden criteria of people who love, dream, believe, deceive and so forth?

The way we answer these kinds of questions will depend upon our view of falsehood. It is not fair to say that in the absence of God there is no truth, no good and no evil, for the humanist and the theist share a conviction about the priority of the values of the living over the phenomena of the inanimate. It is fair to say that without such a value-system *there is no falsehood* (and therefore no truth). This seems surprising because it must surely remain the case that, for example, a theorem is a theorem in mathematics? Yet this is not so, for an invalid proof is as much a phenomenon in the world as a valid one; it is only on the basis of higher values that the valid proof is preferred, i.e. the values arising from the living mathematician. To deny this is necessarily to accept that when a student does something "wrong" in a proof he is not in fact doing something wrong, merely something different. Ascription of truth and falsehood always involves inherent values which are only intelligible from the perspective of the priority of the human or divine. If you reply that a machine programmed to check proofs and separate valid ones from invalid ones will continue to do so, you are correct; but all it will be doing is obeying a set of rules which ascribe to certain kinds of procedure the value "T" (for true) and "F" (for false); neither procedure will be any better or truer than the other.

The second point about falsehood concerns levels of truth and falsehood. It is a complex issue, and I shall only outline it here. Syntactic truth seems straightforward: we imagine that we can recognise a correctly formed statement, and machines can be used to check formalised languages for syntax errors (computers do this all the time). However, sometimes we deliberately break the rules for effect: "this was the most unkindest cut of all", and we do not suppose that in context such rule-breaking is false. Semantic truth is more difficult: someone may lie deliberately, may be inarticulate, or may not be absolutely sure what he wants to say ("I am searching for the right word", he says). The problem is that we cannot express in words in what respect his utterances (formal acts) fall short of truth in the latter case, since to do so would be to relieve him of his problem. Therefore even semantics relies upon our sense of *rightness*, of having "found the right words", and that sense is more than we can express in words. As we proceed through referential, moral and ontological truths the matter becomes more and more difficult, more dependent upon our feelings and convictions, and less expressible in words.

Once we have given expression to our interpretation of the world our formalisations cease to be under our control; they take on a life of their own. Words that can be made to represent the truth can also be made to lie, and creative acts carry with them the potential for abuse. They also carry the potential to be extended and clarified, refined and polished to a degree of precision which was absent from them originally. Words written down as part of a living tradition and on the assumption of a living system of interpretation can assume an authority of their own, so that the question whether there are alternative interpretations becomes of great importance: which interpretation is true? (I am thinking especially of biblical interpretation.) How are we to decide which is the correct interpretation if the texts cannot interpret themselves (as they cannot)?

We can also systematically remove errors from our work by what we have called retrospective refinement, and in so doing we can easily make our ideas less accessible: roughness, groping and fumbling are human; perfection is sometimes cold and intimidating by comparison. We can observe this process in the formation and evolution of mathematics. What begins as an insight into the cross-section of a fallen tree may be developed into a rudimentary roller-concept, from there to the wheel, from the wheel to the more abstract circle, and from the circle to interactions with other abstractions such as straight lines. The fascination of the shapes on flat surfaces gives rise to the mathematics of plane geometry, from which it is noticed that certain arguments and properties recur again and again, and that circles can be conceived in terms of points which are all the same distance from a fixed point. From there a genius such as Euclid may be prompted to set down some fundamental properties of lines and points and angles, and to show that many of the hitherto disconnected properties of other figures can be deduced from them. From the success of that enterprise it is then thought desirable that other mathematical subjects should similarly be reduced to minimal elementary principles, and rudimentary axiomatics is born. What nobody has noticed during this extraordinary progression is that the circle has ceased to bear any real resemblance to the wheel, and that in the strictest terms it has become untrue that a wheel is circular. The circle has assumed a precision which no wheelwright could reproduce; indeed, no circular object produced in any age could reproduce this abstract circle; in physical terms there are no circular objects in the world; in physical terms there are no circles. Perfectionism has again led to inversion, but the inversion (as is so often the case) derives its power from being mistaken for the reality from which it sprang. (The same process occurs as behaviour which arises from the practice of a living faith becomes ritualised and formalised to the point where it no longer bears any resemblance to the lives of contemporaries, but takes on a religious significance of its own we call "tradition".)

These examples illustrate a process which occurs in most fields of enquiry and activity, and which generates a double conceptual system by which we both compensate for and compare the precision of our man-made concepts with the imprecision and uniqueness all around us in the world (the made and the found). In practice no engineer relies upon mathematical precision because he knows that the world is not as the mathematician models it, nor natural laws as the physicist formalises them. For example, every schoolboy learns about Ohm's Law, which relates voltage (V), current (I) and resistance (R) in D.C. electrical circuits according to the equation $V = I \times R$. But no conducting medium obeys this law over the whole range of possible values of V, I and R. A well-known text-book on electrical theory expresses the matter thus:

> Most components in nature are nonlinear. In order to deal with them analytically, however, we sometimes sacrifice some accuracy by approximating the curves by linear segments and deal, then, with linear elements. What we lose in accuracy we gain in predictability. Analytically we can usually predict the operation of the substitute linear circuit with the answer agreeing within a few percent of the actual answer, as obtained experimentally, for the nonlinear circuit.
>
> <div style="text-align:right">Korneff: <i>Introduction to Electronics.</i></div>

But what has happened to truth in this scramble for utility? Does the engineer simply wash his hands of any hope of genuine knowledge of the world, exclaim "we need television sets, not philosophy and theology!", and rush off to design his next circuit or bridge? There is some truth in that, for our inability to handle the truth, when coupled with our needs, forces us back upon expediency. In order to make a mathematical model of the solar system whose equations we could then solve we would need to make so many simplifications that we would have no faith in a precise confirmation of our prediction because we would know that the model we had built was incorrect according to our own principles. In fact, the more often our prediction proved precisely correct the more alarmed we would become, and the less confidence we would have in Newton. Accordingly, we build models which we know are inaccurate, and we rely upon the approximations we make to justify any reasonable errors in the observed values (which we put down to a mixture of theoretical and experimental inaccuracies). In another field, a medical statistician friend of mine says that he can always tell data that have been "cooked" because they always fit the predicted or hoped-for result far too well.

Once theory becomes sufficiently sophisticated to provide a world-view it changes our view of the world. The crucial question then is whether there is any stimulus from the real world strong enough to dislodge the hold of that system on our minds. Critical thought has generated a vast, powerful and potentially all-embracing world-view which although mistaken and destructive has succeeded in reshaping the world in accordance with itself, thus making true what Kant believed to be true, that we make the world intelligible by means of the theoretical structures we impose upon it. There are only two sources of escape from this imprisonment: either we must become aware of some internal inconsistency or incoherence in our theory, or our theory must be challenged by a stimulus of sufficient power to force its reformation. We saw in our discussion of reason that the challenge to change easily evokes fear and resentment as we face the prospect of an unknown and only partly foreseeable future. The challenge must therefore be powerful without being intimidating; it must combine in a finely balanced way an insistence upon the inadequacy of the way we are and a palpable concern for the way we could be which is free of the suspicion that there is some ulterior motive, some vested interest on the part of the challenger which is not made manifest. The address from beyond ourselves and beyond our present experience must evoke our interest, inspire our allegiance, and display its love. One of the most graphic examples of the spontaneous perception of truth is the story of Peter jumping out of the boat to walk to Jesus in the midst of the sea. Any doubts he had initially were overwhelmed by the impetuous decision which so characterised his actions; only subsequently did doubts creep in when the first flush of enthusiasm began to wane. Yet this is exactly the kind of spontaneity and confidence in our intuitions we need if we are to seize opportunities which present us with meaning and truth beyond anything that can be achieved in the safety of the boat.

THE RECOGNITION OF TRUTH

At the beginning of the chapter I observed that we delude ourselves if we imagine that we would recognise an answer to a question whenever we were offered one. Knowing what truth is does not involve knowing how to recognise it. It is all very well to speak of truth in terms of a relationship of fundamental faithfulness, but the possibility of the lie forces us to ask how we recognise such a relationship. How will we recognise truth when we encounter it?

Take as an example the story of Peter jumping over the side of the boat to walk to Jesus. It is reasonable to understand what someone means when he asks "is it true?" in these terms: had I been sitting in the boat with Peter, and assuming that my eyes were playing no tricks, would I have seen Jesus physically walking on the water, and Peter jumping over the side of the boat to go to meet him? Most of us would be prepared to accept that an affirmative answer to this second question would entail an affirmative answer to the first, that is that the story is true in the strongest possible sense. We would also accept in general that there are other senses in which a story might be regarded as "true" even when this kind of condition did not apply. For example, we would accept that there is something "true" about the story of Hamlet, Prince of Denmark, without wishing to impose so severe a historical condition on it because we recognise that the story sheds light on the human condition. But we would distinguish quite clearly in our minds between the former kind of truth pertaining to Peter, and the latter pertaining to Hamlet: the former is truth of a created thing; the latter a truth of a creation of a created thing.

It is characteristic of modern sceptical theology that it is prepared to embrace the second kind of truth but is scandalised by the first. Accordingly, many theologians and some philosophers have tried to rid historical claims of their scandalous element by arguing that all that matters about a story is its meaning, not its historicity. By this they mean that since the story of Peter has the power to convey meaning to us (since it can evoke a deep response in us), as a story it can be regarded as being in the same category as "Hamlet", and the scandal of historicity disappears (where I am using "historicity" in the sense associated with the kind of criterion we originally used of the story of Peter above, i.e. that at a certain place and at a certain time in the physical history of the world certain things occurred which we would have been able to see, touch, smell, hear etc. had we been suitably placed).

This attempt to rid Christian, or other historical claims, of their scandal should be regarded with great suspicion, for it is a device which clearly and unashamedly sacrifices at least one category of truth for no other reason than to satisfy an empirical difficulty we have in one of our current theories of knowledge. Theologians, surrounded by demands to justify the claims of their faith in ways which will satisfy modern sceptical analysis, have taken the easy way out and diluted their teaching in order to make it more palatable. In so doing they have surrendered one of the features of the past which enables it to challenge the complacency of the present, that feature which presents us with claims which refuse to fit into our assumptions about the way the world is. They have reinforced the self-centred rationalism which, in Newbigin's words, "questions the texts without itself being questioned by them". They have colluded with our epistemological self-righteousness, and abandoned one of the corner-stones of our ability to speak with an independent voice.

Central to this abdication is the view that since it is impossible to prove or disprove, verify or falsify claims about the past, one has lost nothing if one abandons the attempt to make truth-claims about supposed historical facts. This is a seductive argument especially to anyone concerned to advocate the Christian position who encounters opposition to it, even genuine difficulty in believing it, for precisely the reason that it makes truth-claims about historical facts. It is a temptation fiercely to be resisted, but one which it is also important to understand. There are several reasons why we should be reluctant to abandon a realist interpretation of Christian stories.

1 — If something is the case we betray our obligation to it if we pretend that it is not; but this assumes that we regard faithfulness to the truth as more important than the availability of scientific explanation. It is to acknowledge that the other takes priority over the self, and that the other has a claim on me which is more important than my desire for proof. The immediate objection to this is that it seems to imply that we should believe everything ever claimed as true regardless of whether we have evidence for it or not (including the claims of witchcraft and astrology). This objection can only be answered in terms of the relation of truth to the recognition of truth, that is in terms of understanding.

2 — An historical story acted out in the world we inhabit has a claim to be taken seriously, especially when it presents itself as a superior way of life, in a way that a story invented by man does not. Whatever meanings we may discern in Hamlet are all coloured by the fact that we know, in the last analysis, that it is only a story, and can immunise ourselves against its insights, if we so wish, by reminding ourselves of that fact. The Church Fathers fought this battle against those who taught that Jesus only seemed to be human, but in fact retained the immunity of his Divinity from pain and suffering, temptation and fear (the Docetists). With a slightly different emphasis from ours, but equally pertinently, the early theologians replied that the unassumed is the unredeemed, i.e. if Jesus was not fully and completely and truly a man like us, then men like us could not be redeemed by him. Part of the challenge of the Gospel to us all is that it arose from the life of a man like each of us, in the world we inhabit, with all the snares and imperfections we must live with. Sadly, and to their everlasting shame, the churches have disinfected Jesus of the smells that inevitably accompany human existence. As a six-foot tall, blonde, blue-eyed,

pale-skinned, handsome man he bears little resemblance to most of us, and we find the observation that he had to eat and drink, wash, and go to the lavatory, rather distasteful, as if it was not "quite nice" to think of Jesus having to do such things. But to distance Jesus from the everyday world is to distance him from everyone he came to save.

3 — The unusual has the power to challenge our complacency about having understood all things, about having supplied ourselves with a clear and infallible criterion of demarcation between the things that can happen in the universe and those that cannot. This criterion is usually based upon an estimate of the methods of the natural sciences and mathematics and the scope and exclusiveness of the theories they have generated. The question we should be asking is not whether science can explain the truth of *everything*, but whether it can explain the *truth* of anything; is it, perhaps, just a hugely successful approximation or convention, as many of the greatest scientists and mathematicians have believed?

4 — Just as truth is not contingent upon proof, so it is not contingent upon empirical verification. To suppose that because we can conceive of no way of providing evidence for historical claims other than those in formal records it is irrelevant whether they are historical or not is to exhibit a mixture of positivism and utilitarianism which involves renunciation of any idea of truth as faithfulness to the other, coupled with renunciation of the uniquely human.

It emerges from these four accounts that the problem of the recognition of truth, far from being solved or eased by the adoption of reason as its sole criterion, has actually been made more difficult. We now tend to rely upon reason and scientific method (as a component of reason) to supply all our criteria of truth and falsehood, despite the severe limitations we have noted. As soon as the reliability of reason is called into question, we find ourselves faced with a yawning chasm of uncertainty which we cannot bridge. If reason falls, then everything falls with it. We attempt to avert this disaster by pretending to ourselves that the domain of reason is not limited, and we reinforce this pretence by systematically eliminating from the category of knowledge any and every claim which, by being inaccessible to reason, would tend to undermine its authority. A method which was designed to free us from all prejudice by providing a clear means of separating truth from falsehood falls victim to the distortions necessary to preserve the illusion of its own omnicompetence.

The problem which must be solved, therefore, is how having increased the domain of knowledge beyond that accessible only to reason, we are to avoid epistemological anarchy (making knowledge anything anyone claims to know); how, having identified the limitations of proof in the establishment of truth, we are to avoid being at the mercy of fallible intuition and even mere whim. How are we to reassure those who see clarification of the limitations of science as the first stage in a process that will open once again all the irrational speculations that science has done so much to lay to rest? This again depends upon the establishment of a coherent account of the relationship of truth to the recognition of truth, that is, to a theory of understanding.

LIVING UNDERSTANDING

AT various points in the last three chapters I have referred to problems which arise from excessive emphasis upon the formal, and in particular from attempts to make the impersonal sufficient for understanding. In fact, whatever we regard as a fully satisfactory life, none of the following formal achievements and actions seems adequate to guarantee it:

(a) allegiance to a particular religious denomination;
(b) educational opportunity;
(c) choice of a particular career;
(d) birth into a particular culture or tradition;
(e) possession of qualifications;
(f) profession of a particular political or religious creed;
(g) possession of particular amounts of money, status, power, fame;
(h) enjoyment of mental and physical health.

If none of these things is sufficient to *guarantee* true life, is anything? Within a formalist world-view the answer must be that where one kind of formal achievement fails another must be substituted: if your religion doesn't satisfy you, try another one; if your job doesn't reward you, try another one; if your wife or husband doesn't please you, try another one. So it goes on. The one thing that is not questioned is *the relationship between the formal and personal worlds*, because that relationship involves issues which we are reluctant to face, issues bearing directly upon our self-understanding.

This view of the formal, which arises from a depersonalised world-view, discounts the contribution of the person involved to the success of the enterprise. This reflection of a reductive and mechanistic world-view discourages us from believing that we can make a significant contribution to the success of our own lives. I do not mean that we do not appreciate the importance of hard work (although that would be true of some), but that we do not appreciate the importance of *personal resolve and determination*. Suppose, for example, someone complains that church services "mean nothing to him", that he "gets nothing out of them". This may, of course, simply be because they are badly conducted or just dull. But it may also be because he does not appreciate the importance of his own contribution to them. Fulfilment and understanding are supposed to be like a jack-in-the-box: open the lid and they will jump out at you; sit passively and wait and they will come to you. But that is not the case, any more than the universe discloses its secrets readily.

We tell ourselves that things that are important ought to be self-evidently so, as if it ought not to be necessary for us to make any personal investment in something before it discloses its secrets. What can we learn from this "deficiency" in the world? The theological imperative with which we began our discussion of reason supplies the answer: the world does not wear its heart on its sleeve because it is important for us to learn how and where to search for fulfilment and understanding, and how to distinguish valuable from valueless pursuits.

Again the difference between a personal and impersonal outlook impresses itself upon us: an impersonal demand that what is important in life should be self-evident and accessible without our making any effort or taking any risks leads inevitably to an excessive dependence upon the sufficiency of the formal and the sufficiency of unaided impersonal reason; a personal world-view, by contrast, recognises that if we invest nothing of ourselves in the world we will discern nothing in it. Adults fluent in a language can compensate for the inadequacies of a child's speech, or the linguistic abilities of a foreigner, by virtue of a determination to understand what they mean. We can compensate for the inadequacies of the formal in order to gain access to the meanings it is intended to convey. Otherwise a vicious circle is created by over-dependence which conforms to what we earlier called the Protection Racket Principle. Life, let us say, is unsatisfying, full of difficulties and problems we cannot solve. Consequently we lack personal resolution and confidence. A formal system, marketed as embracing all known human ills (Marxism, fundamentalism, sacramentalism, creationism, etc.) tempts hard-pressed men and women to reduce still further their self-reliance by adopting it as a creed.

Because of this insufficiency I may learn a formula by rote without understanding the meaning of any of its terms, and rely upon the authority of physicists in affirming that it is true, but I will be unable to use that formula or to draw conclusions from it or to appreciate when and where it is appropriate to apply it. Formal knowledge can be acquired by mechanical learning, but understanding must come from deep within oneself. The roots of understanding burrow into areas apparently unconnected with the subject in hand. The kind of knowledge which can tell whether something is true with or without a proof of its truth or other authority for the assessment cannot be acquired only by formal argument; we must expose ourselves to the reality in question and allow its nature to disclose itself to us so that we begin to feel what it is like to resonate with it. But where do the *determination* and the *motivation* to acquire that understanding come from?

Understanding a language involves more than memorising a set of ready-made sentences: it involves the ability to construct new sentences of our own, and to comprehend new sentences as we encounter them in conversation. Whether I understand the word "table" is therefore reflected by the extent to which I am able to use it or appreciate its use in novel circumstances. The same applies in mathematics: whether or not and to what extent you understand the concept "circle" is assessed not simply by how many theorems and proofs you can quote, but by the facility you have for applying your knowledge in new situations. Theological, mathematical, linguistic and life skills are therefore different kinds of *human* skills which depend upon our understanding. The decisions we come to as a result of applying those skills bear upon our persons, not simply our reason, and therefore our whole person must be engaged in reaching them. I discern whether you understand a word by the way you use it; why should I not discern whether you understand life by the way you live it?

FORMAL AMBIGUITY

When a theologian is asked whether he believes in God he is likely to reply "it depends what you mean by 'God'". This answer can seem like a device designed to avoid the question, but it in fact reflects a general and extremely difficult problem we have with language and communication. It arises in mathematics, where "true" has always to be qualified with "in such-and-such an interpretation". Certain results follow from certain premises, but if those premises are not agreed upon the result cannot have universal scope. From initially confused formal systems we gradually distil economic foundations upon which to build. Refinement of such systems in turn clarifies our understanding by disclosing contradictions, gaps and errors in our reasoning, and suggests other systems which might arise by altering one or more axioms. The theologian is aware that we tend to work with differing systems of interpretation, and that to give an unequivocal "yes" or "no" to a question is potentially to mislead. Does it then not matter whether "God exists" is true? Of course it does, but what that means depends upon what we understand by and feel about God, which in turn relates not to a three-letter word, but to a concept and ultimately to a person, if our particular version of God is real. To believe in a proposition as an intellectual exercise does not necessarily involve any understanding at all of its content, or the existence of that to which it supposedly refers. Sometimes, even when all the formal components are present, it is still legitimate to question whether there is any genuine understanding underlying them (as when a pianist plays a piece without mistakes, yet fails somehow to produce anything musical, or when someone tries to engage us in conversation about the properties of round squares).

I mentioned in the introduction that the possibility of theological conversation depends upon the existence of formal terms referring to a common understanding. The story about the theologian and "do you believe in God?" characterises the dilemma. In a world where there are so many Christian theologies, and consequently so many uses of the word "God", it is perfectly proper to stop and ask which version one is being asked about. It is also proper to ask why one is being asked, for the answer may prove revealing. "Do you believe in God?" is often a euphemism for "Are you one of us?" (where "us" means some sect, fundamentalist group, creationist group, etc.).

What is a man or a woman? A body? An eater and drinker and sleeper? A worker? A soul? A man or a woman is a physical body whose function is the generation of a system of understanding, an orientation towards the world, a mind. At every stage on life's way we display different aspects of understanding ranging from the innocence of childhood, when everything is wonderful and new, to the depth of the wisdom that can come in old age, when everything is at once both new and old. Therefore nothing is conceivably of greater importance than that we should be aware of the systems of thought which are influencing us, for those systems have the power to shape our understanding, and by shaping our understanding to change and mould us as people. However, no one "understanding of the world" has a monopoly, and no understanding of the world is either wholly explicit or entirely self-evident. Therefore if truth can only be discerned within a system of understanding, by what means are we to decide whether a system of understanding is true?

THE PROBLEM OF THE ONE AND THE MANY

During the twentieth century we have become much more aware of the richness and diversity of world religion.

This has brought its own problems for Christians, who often feel that they should be less confident about the uniqueness of the way shown them by Jesus because it seems intolerant to reject what one knows so little about. Yet to explore other religions seems like a betrayal of Christianity unless we learn that Christianity exists for Man, not Man for Christianity. (One thing which prevents us from realising this is that the institutional churches require so much financial support if they are to sustain themselves that it often seems as if their members exist to serve them rather than the other way round.)

The question why we should choose Christianity when there are so many religions to choose from can be illuminated by analogy with a situation in mathematics. There are various branches of mathematics, and within each branch there are further divisions. Different sets of assumptions lead to different sets of conclusions and different spheres of application. Anyone who wants to apply mathematics to a problem must therefore begin by asking what tools he requires, i.e. what kind of problem he is solving, and what a solution to that problem involves. He will then choose the most suitable branch of mathematics for his purpose. In this respect he echoes Einstein's observation that all we need do to apply Euclid to the real world is to add the assertion that the real world satisfies Euclid's axioms, or, to be more pragmatic about it, that in our limited sphere of application some such assumption will do perfectly well. Otherwise we fall into the trap of taking a sledge-hammer to crack a walnut; I do not need tensor calculus to measure the size of a piece of paper because I do not need to take account of the curvature of space-time.

As we remarked in the previous chapter on truth, this scramble for utility seems to succeed only by abandoning truth altogether. But two almost contradictory things can be said arising from it:

  1 — we know from the approximations we are required to make wherever we attempt to make mathematical models of the real

world *that those models cannot possibly be perfect*, and therefore in a strong sense that they cannot be *true*;

2 — we know that despite the imperfections built into our models they afford us an extremely precise predictive power which seems to deny their imperfection.

Successive improvements to our models have brought more and more phenomena within the predictive range of our mathematical science, but the gulf between our models and the world is still unimaginably great.

Mathematics nevertheless has another side to it in which the inverse pair of contradictions seems to emerge:

3 — we know that by limiting our claims to certainty and truth to the properties of the mathematical structures we have made for ourselves we can achieve absolute perfection (for example in the number system, where two plus two equals four);

4 — we know that despite this perfection we must forever eschew any attempt to say anything about the real world as a result of these findings, since such application inevitably requires the perfection of our mathematical structures to be qualified.

In fact, just as a mechanic uses a certain tool for a certain job, so physicists, biologists, economists and sociologists use certain kinds of mathematical tools for their jobs.

This all seems to make the dilemma greater, for it seems to imply that religions should be chosen on grounds of utility rather than on grounds of truth, whereas it seems to be essential that any religion worthy of the name be held to be true. Yet once any religion is held to be true the problem of all other religions being false arises once again, and it seems that to dismiss all the insights of Hindus, Buddhists, Moslems, Jews, Sikhs and Confucianists with a sweep of the arm is to exhibit the very worst kind of religious prejudice. On the other hand, to embrace those insights seems to involve relativising Christianity and abandoning any sense of truth worthy of the name. Are we back to *tertium non datur*?

Humility in science arises from perceiving and accepting the four constraints mentioned above about the extraordinary relationship between mathematics and reality; humility in theology arises from perceiving precisely similar constraints imposed upon us by the relationship between theology and divine reality. For, to rephrase the first two contrasts drawn above in the scientific context, the remarkable thing is that despite the fact that we do not, cannot, and know we cannot, model reality *precisely*, nevertheless we remain rightly confident that the knowledge our models confer upon us *is real knowledge*. Similarly, despite the fact that we do not, cannot, and know that we cannot, speak of God *appropriately* and precisely, nevertheless we remain rightly confident that the knowledge our speaking confers upon *is real knowledge*. In both cases the redeeming feature of our descriptive performance is our ineffable capacity to compensate for the inadequacies of our own formalisations. It is only when we rely upon the self-sufficiency of the formal that our would-be knowledge crumbles into falsehood.

CHOICE AND TRAGEDY

Despite all that has been said we must still pause to heed the warning that there is no impersonal, objective, certain, fail-safe algorithm that will make our choices for us. As creatures living in the stream of the world we cannot stand outside that stream and look at it as disinterested parties. "Stop the world! I want to get off!" is nevertheless a common and heart-felt cry because we dread making a mistake. And this dread arises from the deepest levels of our awareness; levels which cannot be suppressed no matter how hard we try; levels which tell us that we are creatures *born into time and destined to depart from time*. This moment that I am living will never return, and the choices that I now make can never be remade, for however great the similarities between tomorrow's choices and today's, they can never be the same.

So great is our yearning for the cessation of the inexorable ticking of the clock, that we venerate procedures which present us with semblances of timelessness. In particular we have developed a mathematics which is both devoid of time and which treats time as a symmetric variable capable of increase and decrease; and we have developed science on the premise that experiments are *repeatable*.

Our capacity for abstraction permeates our minds with phantasies of ideal timeless worlds in which by endlessly repeatable trial and error, or endless rational analysis based upon perfect data and perfect assumptions, we can make permanently perfect use of our time by making perfect decisions. We anaesthetise ourselves against the reality of tragedy, that this lost moment and this lost life are *irretrievably* lost, that nothing anyone has ever done or will ever do can alter that fact, unless it be God.

> The Moving Finger writes; and, having writ,
> Moves on: nor all thy Piety nor Wit
> Shall lure it back to cancel half a Line,
> Nor all thy Tears wash out a Word of it.
> Fitzgerald, *Omar Khayyam*.

The unbearable reality of the human situation is that our decisions unavoidably affect the world. If we see someone dying and we leave him to die, then he will die; if we perceive something that must be said and do not say it, then it may never be said. Determinism is a mechanism of consolation which says that nothing we do has any effect, and that we are therefore free from such constraints, such contingency. It teaches that nothing we do can make any difference, and therefore that we can do as we like and everything will turn out the same. For the determinist there is therefore *no tragedy, only pain*. But we are *inescapably free*, "condemned to freedom", as Sartre put it so magnificently, and therefore every choice, moment, and opportunity is pregnant with triumph and tragedy. Mathematical logic and scientific method mislead us by their eternity into imagining that truth can stand aloof from time, and that in knowing the truth we know the a-temporal and eternal.

This is tragedy: that I see you making my mistakes, and

warn you as others warned me, and yet you persist because you think you know better. But this tragedy is not a story, endlessly to be retold and embellished, undone and remade again for edification of the literary mind; this tragedy is history and reality which cannot be shrugged off with a laugh, for it is indelibly marked upon your soul as it is upon mine in my time.

This is tragedy: that the concept of reversibility makes us think that it *cannot* — all of it, all of everything — depend upon *this* moment; that would be absurd, unjust, *unthinkable*, that my chance might be lost so trivially, that it might indeed *already have gone*. Everything returns; all things return. It must be so! Anything else would drive a man mad.

We read the Gospels, and we hear of those called to the banquet who were too busy for one reason or another, and we note their replacements from the highways and the hedges, and we say to ourselves "they missed their chance *this time*". But the Gospel contains no "this time": they missed their chance; the feast was over. Christianity is bound up with time, and consequently offends every sense of justice that I have (a sense nurtured on mathematics, logic and science) when it touches upon such things as these: that in spite of having never had the chance (as the heathen); in spite of my not having fully understood (as the fool); in spite of my having lacked a teacher of sufficient persuasion (as the intellectual); in spite of my having had so many other important things to do at the time (as the businessman); *I have had my chance*, and it may never return.

We come from that which is not, and are made of that which is not; except we be plucked from the earth, and taken to that which is, it is to that which is not that we shall return. There is no injustice here. This is pure logic: dust thou art, and unto dust thou shalt return.

When we countenance or perform an abortion we do not do something which is endlessly revisable in the light of alterations to our opinions due to the development of more sophisticated and persuasive philosophical arguments. We destroy a life. There is nothing hypothetical about that, and no amount of anger or protest by anti-abortionists alters that. The irreversibility of time and the transience of opportunity cannot be erased by righteous indignation. Anger is a symptom of impotence, not a remedy for it.

CHOICE AND NECESSITY

At the beginning of the chapter I made two observations: we like to assume that anything really important will be inevitable; we like to believe that anything worth seeing will be self-evident. The logic of the theological imperative is that we *define ourselves* by the choices we make; we become persons who are functions of the choices they have made, including choices made by default by unquestioning acceptance of the values implicit in a culture. This process can be revised in the sense that we can alter course, but it cannot be reversed owing to the nature of time. Only once we have understood this irreversibility can we understand the Christian concepts of forgiveness and redemption. There is no abstract "me" that can be reconstructed from the fragments of choices I *should have made*; there is only the "me" that *is*.

This introduces a parallel between the Christian doctrine of salvation and *constructivism* in mathematics. Constructivists reject appeals to infinite totalities, to objects which could be constructed but which we have not in practice constructed. The radical interpretation of contingency I am advocating here makes the same point about people: never mind what you might have been; consider what you are. This coheres with my rejection of the formal, for it simply will not do to argue that whereas I make profession of a true faith I nevertheless live otherwise; there must also be realisation of the true self. Once that is stressed and we appreciate how far we fall short of it, then and only then can we understand the doctrine of reconciliation. Intellectual recognition of a pattern is not enough; as with the use of words, the adequacy of our personal understanding is conveyed in and revealed by the ways in which we live our lives.

Suppose someone designs a house and produces a step-by-step guide for his builders which will enable them to construct it without further reference to the design. The house is successful, and the algorithm is widely copied by builders who neither meet the designer nor have any idea of his existence. Asked to account for their great success as builders they point to the algorithm, and explain how they do this and this and that, and a house emerges every time. Asked who designed the house they scoff: the house needs no designer; it only needs the step-by-step process which brings it into being. An observer is forced to admit the inevitability of the emergence of the house from the procedures shown to him, and since it is the house that matters rather than whether there ever was a designer, he is content to admit that there may never have been one.

It is indisputable that many complex processes, once thought to be the unique province of human intelligence, have succumbed to automation. The story of the house-builder illustrates three things: the fallacy of *post hoc, ergo propter hoc* (after this, and therefore because of this), that things are not necessarily caused by what precedes them; that it is not necessary to be aware of a design in order to implement it; and conversely that the redundancy of the notion of a design in established processes does not imply there has never been one. The effect of ignoring the design is to reinforce the notion that important events occur with apparent inevitability.

> "Not every one who says to me 'Lord, Lord', shall enter the kingdom of heaven, but he who does the will of my Father who is in heaven. On that day many will say to me 'Lord, Lord, did we not prophesy in your name, and cast out demons in your name, and do many mighty works in your name?' And then I will declare to them, 'I knew you not; depart from me you workers of iniquity.'"
>
> Matthew 7:21–23.

CONCEPTUAL POWER

Scientists base their endeavours upon the assumption that the universe is intelligible, and the systems which they develop and employ are designed to incorporate as much of that intelligibility as possible. But the adequacy of a system does not depend solely upon either its internal coherence or its external correspondence, for neither can ever be demonstrated completely. Instead we are guided by criteria which give expression to a further, deeper assumption about the universe, that its truth and its meaning are inextricably linked together. When we assess an idea, in addition to its internal coherence and external correspondence, we ask questions about its potency, plausibility and potential. Although it is scarcely ever acknowledged, recourse to such criteria tacitly relies upon the assumption that when we discover an idea full of potential, pregnant with meaning, we are right to infer that we are on the track of something important concerning the nature of things. Deep calls to deep.

In mathematics we do not equate quantity with quality. The number of results does not matter in comparison with their richness. Therefore the mathematical "nose", which includes a proper aesthetic sense, as Poincaré often pointed out, is not simply for results which produce legions of other results, but for new concepts and principles which produce results which penetrate into the heart of a subject and lay it bare. It is a particular weakness of utilitarianism as a philosophy that it must be dominated by mass perceptions of utility, and as such is inimical to the pursuit of hunches. Yet epoch-making discoveries and events have seldom if ever arisen from the welling up of mass support; usually they emerge from the painful efforts of some individual fired with a vision for the world.

Powerful ideas and promising avenues of research are seldom self-evident. The mind of the searcher must be tuned to the right frequency before the tell-tale signs can be detected and amplified. But, as with the process of natural selection, all we ever hear about are spectacular successes and failures. With the benefit of hindsight it is all-too-easy to take for granted the self-evidence of results now accepted by the wider community, and to imagine that the pioneers responsible for them never had doubts, or never wondered whether theirs would be the lot of the forgotten explorer. Quite the reverse is the case. The autobiographies and table-talk of such men and women show that they were constantly assailed by doubts, surrounded with problems, and in half a mind to give up. The clarity of the vision which ensues from the completed task gives a false impression of the labour and anguish that accompanied its birth. Often only the slenderest of threads connects the pioneer with his vision. Am I wrong? Is it important? Have I deluded myself? Am I stating the obvious? All these questions are around, reinforced by those ready to condemn the whole enterprise because it does not satisfy their immediate needs. "Why don't you do something more useful?" is the common jibe.

Powerful ideas seem to have a timetable of their own. They make themselves felt at an early stage in their emergence, but prove tantalisingly elusive in the chase. They leave false trails, hide behind disguises, sit infuriatingly on the tip of the mind's tongue and refuse to take shape in a tangible formal expression. They tease by drawing close, flirting with us, to draw us forward, and then disappear leaving us high and dry, exhausted but not victorious. And then, suddenly, usually in the least preposessing of circumstances, they will take shape and come to birth, and we will say "Aha! At last!" But even then their games are not over, for having presented themselves to us they then try to withdraw; we lose the focus that the moment of realisation afforded, and begin to fret lest they should escape us once again. What was clear as crystal begins to grow misty once more, and all the old panics return as we grasp helplessly at shadows and wraiths.

This is one reason why language, formalisation and quotation are so important to us. Although neither a sentence nor even an entire book can express the totality of our understanding and the meanings we wish to convey, they can provide fixed points by which to orientate ourselves. A word, sentence or set of sentences can evoke the pattern of understanding once again, and set off a train of thought. Similarly, when we give voice to or come across a sentence which expresses particularly appositely some aspect of our understanding, we tend to treasure it, and if sufficient people share our appreciation it becomes a proverb or a popular quotation. Famous passages of literature are not famous simply because their author happened upon a particular combination of words, but because those words manage to express in a particularly profound way some aspect of the human condition. They lead us towards the "higher ground", as Robin Hodgkin calls it, and enable us to reconstruct aspects of the ephemeral vision which first inspired them.

The non-inevitability of the decisions we make compares with the non-inevitability of discoveries and solutions to problems. Those who follow find it hard to realise how dark the despair and doubt of the pioneer can be because however hard the journey, mathematical, geographical, or spiritual, they know that someone has completed it before them; there is a destination; all their struggles do not lead only into the dark. The path to the higher ground is easier to find when others have been that way before.

The relative security of our journeys as followers obscures the risk-taking and speculation of the pioneer by emphasising logical progression rather than inspired guess. But genuine creativity involves a complex mixture of conformity and rebellion, for to reach the limits of current knowledge and experience we must stand upon the shoulders of giants (as Newton said of Kepler and Galileo), and therefore accept their authority; and to reach beyond those limits we must abandon some of the constraints which that authority would otherwise impose; we must rebel. Excessive conformity (fundamentalism, absolutism) and excessive rebellion (relativism, liberalism) both lead to sterility: one in the imprisonment of the fortress; the other in the barrenness of the wilderness.

Just as a scientific or mathematical pioneer explores a future unknown to anyone, so we each explore an equally unknown personal future for which there are only the most general of impersonal maps (default strategies). We will adopt those impersonal, institutional practices once we lose confidence in ourselves as persons who not only may but must break the general moulds offered us. Otherwise our lives will be limited by the same kinds of constraints which prevent us from solving mathematical problems: we are afraid to launch out into the unknown; our imagination is cramped; our minds are in chains.

This leads to two observations:

> 1 — it is often to our long-term advantage to persevere with processes and ideas which involve increased complexity in the short-term, and which contradict established methods;
> 2 — this increase in complexity, which demands an increase in effort, is a clue to the direction in which we should search to find more powerful concepts.

Polanyi spoke of a *heuristic field* in which the researcher is guided towards his goal (often an ill-defined goal) by his sense for the line of greatest slope, greatest difficulty. By Robin Hodgkin's mountaineering analogy, the best view (the greatest conceptual power) accrues to those who work hardest to climb highest. But, to pursue the analogy a little further, the abstraction of extreme generality can also leave us short of breath in the rarified atmosphere where concrete examples which will allow us to orientate ourselves with respect to more familiar territory are difficult to come-by.

THE POWER OF THE FORMAL WORLD

The finest idea remains but an idea in the absence of some means of communication. Einstein developed the ideas pertaining to Relativity when young; it was his discovery of the Lorentz transformation and its associated mathematics which made it possible for him to express his ideas in a mathematical form.

Sometimes words become over-charged with meaning. It is noticeable, for example, how even the greatest actors find it difficult to perform Hamlet's soliloquy, for no single performance can possibly express the myriad meanings that have been read from and read into those lines. The same is true of the prologue to St John's Gospel. It is a measure of great religious movements that they give expression to something which millions of people in successive generations find resonating with their own experiences. This association of meaning, importance and truth is neither coincidental nor fortuitous. It reflects the correctness of the assumption of the scientist that wherever meaning and fruitfulness are discerned truth and understanding are to be found.

Scientists exploring a new field of enquiry may spend years searching for some means of obtaining a purchase upon it. They will begin from a vague feeling that something is important, and from that they may digress into a more general investigation of a wide field of issues, rather as mountaineers may require several attempts to find a route to a summit. With echoes of T. S. Eliot, the end of all their exploring may be to return to their original hunch but equipped to recognise in it a richness they could not see before. In particular they may be able *to specify the problem* in a clearer way as a result of discovering an appropriate formal system in which to express it.

To pragmatists and utilitarians this seems to be a piece of self-indulgence: fancy spending years in research only to discover a problem; what we need are more solutions, not more problems! But unsolved problems can often only be solved once we have raised our conceptual system to new levels to obtain a better perspective on what they involve. Mathematical problems, for example, will often yield to a powerful technique developed in an apparently unconnected field where they stubbornly resist solution by more conventional methods. Computer programmers will confirm that it is often easier to write a general-purpose program than a highly specialised one. We can only attempt problems using what we regard as possible methods of solution; new problems generate new methods of solution. The pragmatist or utilitarian argues, in effect, that we should never search for new gold-mines until the existing ones are exhausted.

DEMARCATION

Let us try to sketch an answer to the question about witchcraft and astrology we mentioned much earlier. Just as pupils like to think that clear-cut answers are more desirable than general principles, so we like clear-cut criteria of demarcation by which to separate truth from falsehood, sheep from goats. In the presence of such criteria we know where we are. Therefore it comes as a shock and a surprise when we learn that such criteria do not in general exist in quite the form we desire. There is no clear and overwhelming demonstration of the falsehood of astrological claims which will convince everyone, including their most fervent advocates. There is no clear and overwhelming demonstration of the existence (or non-existence) of God. The crystal clarity of mathematical proofs and disproofs is an artificial clarity, for it creates an expectation of standards of decisiveness which few if any real-world situations can mirror.

Does that mean that astrological claims should not or cannot be rejected? On the contrary; it is of the utmost importance that they be rejected, and for the best possible reasons. They should be rejected not because they have been falsified, but because they conflict with a superior understanding of the universe, because astrology gives rise to an inferior system of understanding, and in so-doing generates a defective orientation towards the world which cramps the mind. It must be rejected because it is deficient in meaning and conceptual power.

What prevents this from being nothing but a prejudice? In a sense, nothing; and in a sense, everything. There is every reason not to call a belief or unbelief a prejudice once we understand that we cannot avoid making decisions, and that all decisions are based upon what are in the last analysis inadequate foundations and data according to formalism and rationalism. But *all* decisions, however humble, are indefensible on such grounds. While we cling to an implicit atomism, and search for the unmoving ground of certainty or the indivisible minima of knowledge somewhere below us, at what we imagine to be a more fundamental level than our own, we will be unable to escape from the conclusion that in the absence of such foundational propositions or particles (an inevitable absence, since there are none) *all* knowledge is prejudiced and uncertain. But my rejection of astrology, witchcraft and the like is not uncertain: I reject them universally, and with absolute confidence and determination. I do not do so because I have access to evidence which makes me confident, or recourse to an argument which makes me certain; I do so because my confidence in the world-view that I already partially understand is great enough to show their palpable absurdity. Conversely, my confidence in the essentials of Christian teaching arises from a corresponding sense of the fruitfulness and potency of that system of ideas as a means of understanding the world and my place within it which all arise from a supervening vision of the Truth. It is not based upon evidence, whether biblical or experiential, but upon the scope and integrity of Christianity as a whole and the persuasiveness of the person of Jesus who stands in the midst of it.

MUTUAL AUTHORITY

Consider two cultures, one from the distant past believing in gods and demons, the other our own. To discount the primitive metaphysics *merely* because it looks different from our own is to fall into a version of formalism, mistaking the expression for the meaning. The first thing we must do is to ask how this primitive people understood the world, which will involve us in what Richard Rorty calls a *restorative hermeneutic* (in *Philosophy and the*

Mirror of Nature), that is an interpretative method which salvages the legitimate insights of other cultures. Such a culture may be capable of challenging our world-view by presenting us with new ways of looking at the world, albeit expressed in terms we find scientifically untenable. What then becomes possible is a *conversation* between two world-views which does not require a common conceptual matrix because hermeneutic skills enable the protagonists to discern the others' meanings, and to respect one another as mutually authoritative.

The possibility of conversation only breaks down when protagonists *do not know what their position is*, and not, contrary to the relativists, when they pretend that they are both as right as one another. But we cannot know what a position is when those who hold it allow their formal systems to be diluted and confused by an insufficiently precise usage, or by permitting a range of views within the boundaries of their position which make it impossible to define. Conceptual clarity is not important because people need to feel secure (that is absolutism at its worst); it is necessary so that conversation can take place with those who hold other positions. The strength and importance of mathematics do not arise from its hold upon reality, or from its criteria of truth and proof, but from its *conceptual clarity and power*.

In mathematics, incompetence excepted, we usually know what we are talking about, but not because mathematical formalism is precise; we know what we are talking about because the dual control exercised by mathematical formalism and mathematical concepts yields a conceptual clarity which enables us to distinguish fruitful ideas from worthless ones and true assertions from false ones (certain hypotheses excepted). *There is no reason why the same should not be true of Christian theology*. But for it to be true of Christian theology the Church must give much clearer expression to its beliefs.

Good formal mathematics arises from a firm grasp of mathematical concepts, that is from mathematical understanding. Good formal theology arises from a firm grasp of theological concepts, that is from theological understanding. Just as no teacher will be fooled into believing that a pupil "really" understands, despite manifest formal incompetence, so no-one should be fooled into believing that a church "really" knows what it believes despite manifest formal inconsistency. This is *not*, let me emphasise again, a matter of inquisitorial suppression of divergent views; it is a matter of conceptual clarity or, in St Paul's phraseology, spiritual discernment. It is a matter of having the mind of Christ.

THE RATCHET PRINCIPLE

Current theories of evolution have managed to take the sting out of the common criticism that the emergence of life by purely random mutations is inconceivably improbable within the time we know it to have taken. They have done this by observing, and demonstrating experimentally, that if certain configurations are more stable than others among a random distribution of events, those configurations will tend to become predominant in any population. As a trivial but accurate example, suppose that we throw a fair die until it comes up six, then place it on a chess-board, and then throw another die until it comes up six and do the same on an adjacent square. Soon — remarkably soon, in fact — the chess-board will be full of dice showing sixes. If we invite a stranger into the room and ask him to calculate the probability of this arrangement happening purely by chance, he will find that it is inconceivably improbable (in fact of the order of 6^{-64}). He will be tempted to conclude that something has happened other than pure chance (and he will be correct to do so). In fact, as we know, "six" was a stable, surviving configuration, and all other numbers decayed very quickly. Evolutionary biology now relies upon this kind of account to refute the probabilistic criticism mentioned, which has been put forward by some eminent mathematicians, including Kurt Gödel the great logician.

Translated into the sphere of human understanding

this same ratchet principle can be used to illustrate the asymmetry of the growth of human knowledge. As we saw very clearly in the first chapter, mathematicians find it hard to tell what they know; students suffer because their teachers are generally unable to tell them how to proceed from non-understanding to comprehension. A teacher can *show* them how to tackle a problem, but cannot (other than by taking them step-by-step through it) *tell* them how to do it. This is because the teacher has made the transition himself, but does not understand how he has made it.

Before something becomes clear to us, before we understand, all the various fragments of knowledge, hints of connections, misinformation and confusion clatter about in our minds making no sense at all. Then something "clicks": we understand, and the fragments suddenly disappear. Not only do they disappear: we cannot even say what they were; it becomes almost impossible to understand how we could ever have *failed* to understand. The item has become *obvious*. What is more, having hitherto identified fully and passionately with our fellows who were also struggling, we now do not understand their struggle: we can assent to it formally, and see it, but we cannot *feel* it; it has ceased to be a part of our being. Obviously our problem as would-be *teachers* is that our task is to enable our students to move from fragmentation to integration, to understanding; but we cannot account for this transition in ourselves, or identify very readily with the fragmentation they experience.

The same thing happens in a mathematics problem. First of all the solution is unknown, then we struggle with various possible solutions for a period between seconds and months, and suddenly the solution becomes clear. Thereafter (barring extreme forgetfulness) the *right* path looks quite different from all the wrong paths, almost as if it was illuminated differently; but for the student this differentiation is absent; the correct route to a solution looks just the same as all the incorrect ones. we are able, of course, to see these wrong solutions as teachers. They appear again and again in work that we mark. What we cannot do is, so to speak, to *blot out* the illumination of the right one, to make it once again seem just like all the others. We cannot selectively and deliberately *forget* what has become clear.

Consider a more familiar example. Think back to when you learned to ride a bicycle. The transition from incompetence to competence was probably fairly swift, a matter of hours or days. Now — can you put yourself in the position of someone who cannot ride one? Can you even begin to imagine what it is like *not* to be able to ride a bicycle? If you tried, and sat on one would you not need to *force* yourself to fall off? Otherwise, would it not be inconceivable that you would fall off?

One morning my daughter, aged sixteen months, stood up in the middle of the dining-room and walked six steps. No power on earth could have forced her to take that monumental, and yet in some ways inconsequential step into the world. The peculiar combination of things going click inside her that made it possible, a mixture of desire and ability and confidence and belief in herself, simply had to occur. She had to decide that walking was a good thing to do; some kind of inarticulately perceived possibility had to seize her mind.

Mathematics and Christian faith share this feature: once we understand parts of either of them it becomes virtually impossible to put ourselves in the place of someone who does not; when we do not understand either of them we find it impossible to see what that understanding that we lack could involve, or even why it matters that we should acquire it. From outside looking in both seem irrelevant and obscure; from inside looking out they appear to offer us the keys to all things. Any attempt to give an account of mathematics or Christian faith *without* the dimension of integrated understanding would be inadequate: what matters is not the theorem or the proof, but the understanding of the theorem which makes the proof dispensable; what matters about the faith is not its dogmas or their justification, but the understanding of those dogmas which renders the justifications unnecessary.

Yet it is precisely these "understandings" which we cannot put into words; they are part of the non-formal world which governs and is enriched by the formal world without being reducible to it. It is as if I need proofs and justifications as ladders to climb a tree whose trunk lies elsewhere; once I am among the branches, the ladder can be removed without my falling down.

Carnes uses the rather unreal examples of the Necker Cube and the Duckrabbit as illustrations of perceptual problems. But how do you describe an "A" without referring to its A-ness? A geometrical description will not suffice: it tells you only about the relative positions of lines. How do you understand that something is a solution? Or that it is the truth? We see the world, and it makes sense to us at certain levels. But suppose we are as ignorant of the real significance and nature of the world as someone who sees "A" as a set of lines variously disposed is ignorant of its A-ness? We cannot lift ourselves up by our own bootstraps, but we can assent to the possibility that there are higher planes of understanding than our own. Some people understand Jesus as a man; some see in him the self-revelation of God. The second is a self-authenticating vision; nobody would dream of believing it unless they also believed it to be true. And once the greater truth has been seen, how can it be forgotten?

A critic would charge such a position with circularity, but such circularity is inescapable. Since the understanding is what really matters, and since that understanding can only be appreciated in a non-formal way — since it is accessible to emotion and feeling rather than to dispassionate reason — the only way it can be evaluated is by first being experienced as a result of some kind of integration of fragments to their whole. The justifications of a position which are accessible *from below* are of use only to help someone to understanding; they are not substitutes for understanding. The real justification of understanding is accessible only to those who understand, because only those who understand have access to the self-authenticating vision of the truth which makes justification in terms of the "below" and "before" redundant; we then put away childish things.

INTIMATIONS OF FRUITFULNESS

A rationalist still regards this as a scandal: how dare anyone claim that in order to assess a position you must first swallow it "hook, line and sinker"! That makes assessment impossible, for unless one is persuaded of the truth of the position one will not wish to commit oneself; but if one does not commit oneself there can be no hope of assessing its truth.

This is rather like the problem Plato addresses in the *Meno*: to find anything new we must look for it, but to look for it we must already know what we are looking for; therefore either we cannot look for something new, or whatever we find cannot be new. Polanyi replies to the paradox by rejecting one of its assumptions, that in order to look for something we must know what it is we are looking for. He can do this because of his notion of *tacit knowing*, which relies upon the formal — non-formal distinction. By dwelling in something I become focally aware of its meaning; but focal awareness does not mean precise, crystal-clear awareness; it means something closer to a *feel* for its meaning, which in turn gives a clue to its *value*. I can find something new because those aspects of my person which extend beyond pure reason can sense richness of meaning, intimations of fruitfulness long before anything that can be formalised comes into range of my rational powers. It is this ability to *sense* important and rewarding problems which Polanyi praises so highly in a scientist, but it is also of vital importance whenever we make choices in our daily lives, and depends upon our having trained those distinctly human senses which indicate promising avenues of exploration.

PERSONAL CHOICE

What am I to do? What am I to believe? What am I to do with my life? Whom shall I follow? What ought my values to be? These are everyday questions, but they seldom receive satisfactory everyday answers because our *choice* and the *mechanism* of our choice have been severely circumscribed by rationalism, by exactly the assumptions which give the force to the *Meno* paradox, namely that we must have explicit knowledge of an answer before it counts as one. When we began our examination of truth I pointed out that it is an illusion to suppose that we always recognise an answer when we see one. An answer which is more than trivial involves *change,* because it conflicts with our existing explicit knowledge. Rationalism depends upon the existence of explicit, propositional data which it can operate upon. But in most of our critical life-decisions those data are absent or unavailable; we simply do not and cannot know what kind of person we will be when we are forty when we make life-directing decisions at fifteen, twenty, or twenty-five. Therefore, if the decisions (which obviously have to be made) are to be made with all the assistance of all the faculties at our disposal we need a much richer fund of choices and resources to draw upon. In particular we need that rich integration of reason, emotion, belief and feeling which is realised in *persons.*

Fulfilment in life depends upon learning how to tell what and whom to pursue. In some form or other we must each decide upon:

1 — a problem worth pursuing;
2 — a career worth pursuing;
3 — a relationship worth pursuing;
4 — a hunch worth pursuing;
5 — a lifestyle worth pursuing;
6 — a religion worth pursuing;
7 — a personhood worth pursuing.

All these decisions involve learning how to recognise what matters. We only have a finite amount of time to make a finite amount of choices; the clock cannot be wound backwards. How do we decide what to pursue? And what determines the range of choices available?

Societies stand or fall, in the long term, by their success

in creating an environment conducive to the development of responsible citizens capable of making the correct decisions about the course of their lives and their conduct in society. Such citizens act as centres of understanding capable of enriching one another's lives without massive centralisation and direction. Such an environment will not arise simply by provision of mechanical rewards and punishments (the familiar "carrot and stick"); they will only serve to order society. It involves the emergence of an atmosphere of expectation in which understanding is valued despite its incommunicability, in which merely mechanical performance at every level is valued less, and in which the highest accolade is reserved for those who have that rare and extraordinary power to evoke understanding in others.

The most important ingredient in our society is therefore that quality which encourages us all to stretch out beyond ourselves and our self-centredness to achieve contact with the other, whether material, personal or divine. That ingredient is the *love* which finds fulfilment not in selfish manipulation of the other, but by acknowledging its own incompleteness and the richness of the other as a complement to itself and worthy of regard for its own sake.

PERSONAL AUTHORITY

There is, however, another possibility. If I cannot persuade you with arguments that something is important; if you are unimpressed by my attempts to demonstrate its relevance to you; if you do not see what use it is; you may still pick something up from my obvious enthusiasm for it. You may, of course, merely dismiss me as mad or misguided, but if there is something about my interest in mathematics or Christianity which touches you at a human and personal, rather than an intellectual level, you may nevertheless stop and take more notice. That was clearly the case with Jesus: he spoke with authority, with words which emanated from and were consistent with his whole being as the Incarnate Word, and he was heard to be and recognised as such by people who were far from being intellectuals.

Most of what has been said in the chapters on Proof, Reason and Truth is more closely related to Greek philosophy than to Hebrew theology. But the Christian world-view integrates both in the concept of the *logos*, which is partly derived from the Old Testament concept *debhar Elohim* (Word of God), and partly from Jewish-Hellenic and Stoic sources, as C. H. Dodd explains (I have transliterated the Greek words):

> The word *logos* has an extremely extensive range of meanings. Those which must concern us here are the two which the Stoics distinguished as *logos endiathetos* and *logos prophorikos* — the *logos* in the mind and the uttered *logos* — i.e. "thought" and "word". ... *Logos* as "word" is never the mere word as an assemblage of sounds (*phone*) but the word as determined by a meaning and conveying a meaning (*phone semantike*, Aristotle, *De Interp.* 4). *Logos* as "thought" is neither the faculty nor the process of thinking as such, but an articulate unit of thought, capable of intelligent utterance, whether as a single word (*hrema*), a phrase or sentence, or a prolonged discourse, or even a book. ... Behind it lies the idea of that which is rationally ordered, such as "proportion" in mathematics or what we call "law" in nature.
>
> C.H.Dodd, *The Interpretation of the Fourth Gospel* p. 263.

Thus the Greek Stoic element is integrated into the Hebrew tradition in the person of Jesus of Nazareth. The rational principle which creates and governs all things takes flesh in the world and also utters words which are themselves creative. Words become creative by making possible what remains impossible but for their being spoken; they evoke in their hearers a sense of human possibility which otherwise remains unknown and unrealised. It is that call which beckons us away from earthbound security and certainty as typified by objectivism, proof and self-centred reason, to participate in divine rationality by relating to the Truth as person.

For the Stoics, the *logos* was an impersonal metaphysical principle, an idea; in the Christian revelation it makes itself known as personal and interactive. It is what it is, but it is capable of relating to us as we are, of knowing imperfection without becoming imperfect, of taking sin upon itself without becoming sinful. In this it shows the possibility in the human situation, the possibility which we either feel that we cannot achieve or do not recognise as a possibility at all. That possibility is our hope which, not simply as a story or as an edifying discourse but as a matter of fact was realised in Jesus, in our world, in our mess and from the midst of our despair.

The proclamation of the Gospel of Jesus is the proclamation of that hope as possibility, but not only as possibility. It is also a proclamation of facticity, of achievement, of the realisation of the Truth in the midst of the world. That Truth is the living antithesis of earth-bound security, whether it be the security of fundamentalism or sacramentalism, rationalism or irrationalism, sectarianism or institutionalism. That Truth denies the legitimacy of pretending that we can absolve ourselves from responsibility for what we believe and know and do by referring to impersonal facts, authorities and powers. It says that we must be responsible for ourselves, that we must realise in ourselves the possibility which can only ever be presented, not enforced. The realisation of the human possibility is the equivalent, on a life-long scale, of the infant suddenly discovering the ability to stand up and walk, or of Peter uttering the ultimate confession of the faith: "Thou art the Christ".

Understanding is not explicit or formalisable. It cannot be taken out and shown to the world. But it reveals itself in all that we say and do. The attitudes of the mind and heart reveal themselves in our words and actions, either as integrity or hypocrisy. There are finally no partitions or compartments in our minds and lives. It is an illusion to suppose that one can think evil thoughts and live a blameless life, separating the things of the mind from the actions of the body. Philosophically, Polanyi showed us why that is so in his notions of indwelling and tacit integrations: those influences to which we subject ourselves will inescapably shape our lives. Jesus' teaching that to harbour thoughts of adultery and murder is the same thing as to commit them points to the same truth: the way we conduct our lives arises from the thoughts we think and the influences with which we feed ourselves. That is the simplest of resolutions to the perennial and vexed question of the Communion service: to eat Christ's flesh and to drink his blood is to feed ourselves with the Way, the Truth and the Life. The bread and wine are a focus for what must be a lifetime of feeding through participation and relationship. Sacramentalism substitutes a formal practice for living nourishment, and underestimates the force of John 6:63, "It is the spirit that gives life, the flesh is of no avail; the words that I have spoken to you are spirit and life". Institutionalism substitutes mechanical behaviour for community life. The question of understanding leads inexorably to questions of ministry and the body of Christ. The Jesus who is the Way is the personal answer to the unanswerable questions about what is worth pursuing, for in him the formal and non-formal, the word and the understanding, the body and the mind, act in unison, and as such in fullness of life. The Church is charged with the task of continuing to speak in as close a way to that as possible. That will only be possible where the people of the Church are so orientated to God that such an understanding governs their lives. The Church is where men and women of such faith are to be found.

FINITUDE AND CHOICE

During our adolescence we believe that we can do everything: we can become Prime Minister, brilliant surgeon, nuclear physicist, best-selling novelist, England cricket and football captains, etc., etc. Everything is in the future, and since most things in the future are possible we can set aside our finitude and the need to make choices which inevitably close more doors than they open.

In chapter two I set out four challenges which a post-critical philosophy must meet. The discussions in this chapter pinpoint the single problem which makes a generally adequate response difficult. That problem consists in the incompatibility between two philosophical positions, which I shall set down as C (for critical) and P (for post-critical):

> C — There exists an impersonal basis upon which to reach objective decisions about best systems of ideas, best descriptions of the universe, best ethical systems, and it is our duty to adopt such a stance in deciding upon and coming to all our fundamental beliefs.
> P — The only basis upon which we can ever make decisions or reach conclusions about anything is a personal one governed by our cultural heritage, genetic constitution, and environmental experience; our task is to acknowledge the inevitability of the personal and to develop systems of mutual correction and authority which will generate human and personal systems of ideas, science, ethics and fundamental beliefs.

These two positions are mutually exclusive in the strongest possible sense, that questions posed from C to P or from P to C are unintelligible: they are not questions at all (any more than "how many sides has a round square?" is a question at all for someone who understands the meanings of "round" and "square"). From perspective P, C-questions reveal a fundamental failure to understand the nature of the human situation; they arise from a mixture of absolutism and depersonalisation; they presuppose that it is possible to know without being involved in one's knowing. From perspective C, P-questions display subjectivism and prejudice, failure to take sufficient account of the distortions which accompany all personal involvement in knowing, and consequent failure to adopt an impersonal frame of reference which self-consciously seeks to eliminate that frailty.

"How much did you find out about other faiths before deciding upon Christianity?" is a C-question. It presupposes that there is a religionless state from which to assess the relative merits of religions. But from a religionless perspective no religion has any merits. The questioner is really asking about the difference between my Christian faith and his unbelief (or perhaps Moslem or other faith): how is it that you have come to a decision and I have not; how is it that you have decided this way, and I that? The P-answer is that the basis of belief is not formal analysis. I am not a Christian for this or that reason, or even for a number of distinct reasons strung together: Christianity makes sense as a whole, and confirms its truth through its fruitfulness. I can give you many reasons for holding Christian faith, but in the end they will be inadequate unless you appreciate that in the end I do not hold it; *it holds me*. I am *persuaded*, but not in the way that the objectivist claims to be persuaded, by impersonal and compelling evidence which exempts him for responsibility for what he believes. An integration has occurred, the ratchet has turned through one more sprocket, and a stable configuration has arisen which cannot be accounted for solely in terms of what preceded it.

Religion and politics are regarded as subjects which should be excluded from social conversation because they notoriously lead to heated argument. It is a commonplace to hear religion accused of being responsible for more war and human suffering than anything else. The arguments we have pursued in the present work show why that is not true, or at least the limited sense in which it is true. Wars arise when positions upon which we rely for security are under threat, when the props we use to gird up our lives are in danger of being knocked away. To threaten my tribe is to threaten me because my life and identity are made up by my place in the tribe. All these considerations point to our failure to achieve personhood as the root of the evil, not our religious belief. The religious and the unreligious both share a dependence upon impersonal things and standards which prop up their inadequacies as persons, as fully human beings. The greater that dependence, the greater the threat when the props are attacked, and the more violent the response. Someone who has achieved personhood is at one with himself, his world, his neighbour and his God. He has become whole and full. Jesus was such a person. Even in the loneliness of his last hours he retained the wholeness of being which can only arise from those who, depending upon nothing that men can destroy, have nothing to fear from men. That which seeks to fill itself becomes empty, and that which empties itself is filled to overflowing. That which is full to overflowing gives without becoming less and receives without becoming more.

Those who followed Jesus in the flesh two thousand years ago did not understand *intellectually* any more than we do with all our theology, but they perceived that there was someone here more important than life itself, and they followed him long enough to glimpse something of the truth for themselves. They did not know *what truth is*; neither do we. But they understood *where truth was to be found*.

FUNDAMENTALISM

The New Testament is a record of Early Church understanding of the significance of Jesus within which we find attempts to record what he actually said and did. The Early Church knew more than it could tell, and its formal witness means more than it could ever know. Positively, these documents preserved genuine insights of these early Christians, and provided a yardstick against which all subsequent Christian writing and speaking could be measured. But the negative effect of canonising these texts was to introduce a discontinuity into the Church's witness, and to pave the way for biblicism. That is not to deny the extraordinary nature and power of this witness: indeed, it can only be appreciated when compared with the crudity of similar contemporary documents; but the existence of canonical scripture tended to block development of an adequate appreciation of the work of the Holy Spirit which recognised the necessity of his presence in every generation and in every Christian mind if appropriate understanding of those texts was to ensue. Neither the biblical text nor any other text can be called "The Word of God" without undermining this necessity. The Presbyterian phrase "Let us hear the Word of God as it comes to us in ..." is much to be preferred to the coarse Anglican "This is the Word of the Lord", because it emphasises the dynamic living relationship necessary for any human word to become a vehicle for God's Word.

ANALYSIS AND INTUITION

In his essay "Leibniz and Descartes: Proof and Eternal Truths" Ian Hacking draws a sharp contrast between Leibniz as the champion of proof as the road to truth, and Descartes as the champion of intuition. As with so many things, this sharp distinction between proof and innate ideas will not do; it is too neat. Analytic and intuitive approaches to understanding are often set in opposition in such a way, but nobody who has done any mathematics or creative writing is likely to be convinced by this way of describing things.

When we set about solving a problem in mathematics we begin with a certain amount of knowledge about mathematical tools which may or may not be relevant; we also have experience of certain kinds of problems which may be useful; and we have some rather vague skills which can be summed up in terms of a general method of approach. At first we may be uncertain how to proceed, and try several avenues of approach using pencil and paper. We do this because we know that setting things down helps focus our minds, and because the formal (for reasons we discussed in chapter two) has the power to reveal new relationships which might otherwise be missed. More often than not a solution goes through several hiccoughs before it is finished as points of detail require more careful attention, but eventually something like a solution emerges. Some rest content with this, but we usually look at the solution and ask whether it can be improved upon. If the result seems straightforward we may even ask whether there is a simpler approach (as we will do if the algebra or method seems unduly clumsy). If the problem is non-trivial we usually learn something from solving it, and place, so to speak, a further brick on the pillar of our abilities from which to assault greater heights next time.

If we write an essay on theology we start by setting down the broad outline, showing principle divisions of the subject, and the main points we wish to make. We then begin to write, and often the first line takes a long time to compose. When that seems right the speed of composition can increase as we weave our way through our skeleton. But, as in a mathematical proof, the process of writing stimulates further thoughts which force a revision of the plan. Sometimes we will be reminded of work by someone who has made salient points on the topic; reading a few pages of that work may prompt further ideas. The finished product then requires editing and revising, and in the end we should (if it is a good essay) find that we have learned something new in rather the way that the mathematical proof teaches us something about connections lying deep within the subject.

These illustrations confirm our inability to foresee all the implications of a point of departure, the need for some kind of guide if the sheer variety of possibilities is not to swamp any kind of coherent route. But most important of all they indicate the mutual dependence of the vision and its expressive medium: not only must a vision have such a suitable medium available; that medium must be given shape by the vision. The vision acquires depth and detail by virtue of the availability of the means to elaborate upon it; the expressive medium takes shape and assumes an appropriate form by virtue of the guiding vision. An artist strives to achieve truth in all five respects: technically; expressively; referentially; morally; intrinsically. Only when the intrinsic adequacy of his work satisfies him will he rest; sometimes he will destroy works which lesser mortals would be proud of because they fail in this respect.

In mathematics we find that such intrinsic truth often conveys to us an understanding of what is entailed by a theorem which is lost when doubtless adequate tools or proofs of lesser beauty are used: clumsiness clouds our perception. Similarly, the finest theology combines a depth of perception with a beauty of expression. That beauty is often characterised by a deceptive simplicity: we make do with a long and clumsy proof or essay because we have no time to construct a shorter one!

The sense in which the artist, writer, mathematician is the judge of the intrinsic adequacy of his own work is a vital indicator of the relationship between person and world. There are three distinguishable stresses in this relationship which can be summarised as:

> *this* is what I am obliged to say in that I have perceived, however dimly, some aspect of the truth which must be shared;
> this is what *I* am obliged to say because this perception has come perhaps not to me alone, but nonetheless to me and therefore if I do not give it expression it may be lost;
> this is what I am obliged to *say* because no vision, however humble, is a gift solely to the recipient, and therefore must be expressed in some form.

The mutual dependence of analytic and intuitive modes of thought also illustrates and illuminates the dilemma of the student who is trying to internalise the understanding of others in order to begin from where they stop, and to make a contribution of his own. Neither as mathematician nor theologian can I take my understanding out of my head and implant it in someone else's. Moreover, the world seems to be constructed in order to make this as difficult as possible. Simply reading my words or following my proofs will not guarantee the success of this process (any more than pursuit of certain other formal procedures will guarantee fulness of life). But if you try not simply to follow but to understand the processes behind each step of the argument by reintegrating the parts to a new whole of your own you may come to share that understanding. It will not be mine: it will be better than it would be were it merely mine; it will be mine and yours combined in a quite unique way. You may make mistakes, and I may have misled you or simply been mistaken to start with. But the power to correct those mistakes also lies in you.

SUPERVENING VISION

Throughout this work we have come back again and again to the problem of being wrong. In this chapter we have seen some of the pressures from the multiplicity of religions which tempt us to abandon any sense of a unique truth in favour of a tolerant pluralism. In both mathematics and theology we have seen the inadequacy of the merely formal and the dumbness of the merely nonformal. In particular we have seen the logical dilemma of incompatible systems of logic and thought which threaten permanently to divide us from one another. One solution to these divisions gives rise to conflict, physical or psychological; another gives rise to apostasy as to the concept of one universal binding truth. Is there a way through this morass of interwoven problems?

I am convinced that the answer is extremely simple, so simple that it seems almost impossible that it should either be overlooked or ignored. It can be expressed in two paragraphs:

First we must abandon belief in the sufficiency of the formal in science, religion and general life. In particular we must stop believing that salvation comes on the basis of *right formal confession* or *right moral action* (i.e. justification by creed and works are both equally mistaken). Moreover we must include within this prohibition belief in the words of the Bible, belief in the creeds, doctrines and disputations of philosophy and theology, and belief in the rites and practices of the Church. Most important of all we must abandon the conviction that our salvation (in this world or the next) depends on our holding on to a certain system of formal beliefs.

Second we must strive for vision of the Truth which in its absolute integrity binds, unifies, guides and frees all things. "Binds" in that in its utter universality it demands universal assent; "unifies" in that by virtue of its totality it condemns as inadequate modes of relating which involve less than the totality of the person, and as such condemns formalism in all its shapes and disguises; "guides" in that it shows us the way by revealing the value inherent in itself; "frees" in that however fleeting the vision may be, in its unimaginable comprehensiveness it reassures us that it will never pass away, and that it has no need of us to sustain it, but rather that we can rely on it to sustain us.

In practice the most important consequence for religious people will be the rediscovery of the sovereignty of God, by which I mean that we will learn that God can take care of himself. That may sound presumptuously trite, but it is to be understood thus: all our clinging to systems, whether biblical, ecclesiastical, theological or philosophical, amounts in the end to unbelief. We do not believe that God will be there regardless of what we believe (rather, but not exactly, of course, as mathematical truth will be there regardless of the mistakes we make). We do not believe that God has sent forth his Son and saved us. Therefore we insist on hanging on to our beliefs, systems, rituals, institutions, theologies and philosophies *as if our life depended upon them* (which we believe it does). We are afraid to explore the unknown because we are unsure of the reliability of God's promise that heaven and earth will pass away but that his Word will never pass away. If we really believed that, we would not get so cross and protective about our inconsequential formal systems, our doctrines, churches, rituals and creeds, which at best offer us the flimsiest impression of the Truth which embraces heaven and earth.

I do not suggest for one moment that this vision will offer us comfort *for ourselves*, or that as a result all our pain and suffering will vanish. On the contrary, the precise opposite can be guaranteed, that nothing whatever will change *formally*. We will each doubtless sin and suffer, forget and lie as we have always done; we will know death and we will know sorrow, just as we will know joy and laughter, and in our depression we will doubtless lose sight of the vision which once embraced us and stretched our minds beyond the finite and beyond the unimaginable boundaries of our world. But the Truth will remain, and we shall find ourselves persuaded

> that neither death, nor life, nor angels, nor principalities, nor powers, nor things present, nor things to come, nor height, nor depth, nor anything else in all creation, will be able to separate us from the Love of God in Christ Jesus our Lord.
> Romans 8:38, 39.

The distinction which must be kept clear is that between regarding doctrines, rituals, creeds and texts as *clues and pointers* to the truth which are there to be explored, followed, integrated and understood and, for that reason and in the interests of clarity of thought and ease of access, protected from dilution and confusion, and regarding those same formalisations as in some sense sacred in themselves, to be protected for their own sake and in their own right, in which case they cannot but usurp the place of truth for themselves. To see them neither as beliefs to be held for their own sake, nor as passports to heaven, but as signposts, guides and even rungs on a ladder of understanding, is to be freed to have one's eyes opened and one's mind cleared for the experience of that supervening vision of the Truth which, although forever beyond and before, is nevertheless for us and open towards us. Like the exercises of the young pianist, these self-conscious statements, beliefs and affirmations must be internalised, almost as if forgotten, so that the shape of the whole which is life, and light, and joy, and peace, may be discerned.

EPILOGUE

THE modern era in philosophy and mathematics can be dated from Descartes, who introduced scepticism into philosophy in a rigorous way, and invented analytic geometry and in so doing transferred the emphasis in mathematics from aesthetics to analysis. But the most significant aspect of Descartes' influence lay in his legitimation of doubt as a result of his emphasis upon reason. Hitherto the onus had been upon the sceptic to justify his scepticism; subsequently it lay upon the asserter to justify his beliefs. The transition was neither immediate nor universal, but the dominant themes in subsequent philosophy arose from this contraposition to the extent that from it philosophy took its name, "critical" philosophy. From many points of view this transition represented an unqualified gain: the prejudices which forced a recantation from Galileo could always be cited as an example where the new method would have prevented an injustice. But Descartes' method relied upon the abilities of individuals to decide upon things for themselves if it was not to replace one form of authority with another, and in practice considerations of time and intelligence precluded such a possibility. The authority of Church was replaced by the authority of Science, and concepts of heavenly salvation derived from the first found their counterparts in hopes for an earthly salvation provided by the second.

The understanding of proof which Leibniz developed in this climate was based upon indisputable truths and axioms, and deriving from them unsuspected new truths as theorems. Proof seemed to represent a means of progressing from the simple and unassailable to the more complex and important. Mathematical methods offered hope of knowledge based firmly upon foundations which everyone would accept, giving rise by solid procedures of inference to new insights which those who accepted the premises would therefore be bound to accept as well. It looked as if such a method would one day remove the need for barbarisms such as war: all matters could be settled by reasonable men in ways which would command universal assent. No-one doubts that this was a forlorn hope, but the reason for its disappointment does not lie, as is commonly supposed, in the fact that common assent to premises cannot usually be attained. It lies in the fact that the view of mathematical proof inherent in this account is mistaken: the axioms or premises themselves have a history, and their hold upon us arises from our participation in that history, not vice versa. Even if I am persuaded to assent to a set of premises, and acknowledge the correctness of the inferences drawn from them, I will not be persuaded to accept a conclusion which contradicts my world-view unless it is of such striking power that it overwhelms my previous commitment to another system (in other words, if it is sufficiently powerful to *convert* me). In practice this is seldom the case: our prior convictions as to the truth or falsehood of certain statements *over-rule* any proof to the contrary. Thus, for example, no proof of the existence of God will convince an atheist because his commitment to his atheism will lead him to renounce aspects of the proof, usually by contesting a premise or an inference, in order to sustain his stronger atheism.

Reason is always the servant of our deeply held convictions, even when we suppose that we have mastered them, and only by incorporating a proper estimate of those convictions into our educational and rational systems of behaviour can we hope to progress. Part of this solution relies upon perceiving the self-centredness of most reasoning, and the dominance of our fear of error and failure in our culture. Rationalism has failed, and in its failure has brought the human race to the verge of self-extinction, because it supposed that it could first suppress and then ignore the emotions. It was consequently helpless to combat irrationalism, as the pitiful inadequacy of the liberal response to Naziism and Stalinism showed.

By focussing upon repeatable and demonstrable proofs and impersonal reason our cultural disposition led naturally to *formalism* in which words, symbols, rites and actions of an assessable and repeatable kind would be dominant. Because all formalism can by definition be detached from its author, it seemed an ideal depersonalised basis for knowledge in which the rationalist dream of eliminating unpredictable emotional content would be realised. But the distinction between what a man is and what a man does, or between what a man thinks and what a man says, must either be ignored in such a system (with obviously catastrophic results), or remain as a permanent embarrassment to it.

The value of an impersonal, formalistic understanding of truth to a culture which was gradually losing its grip upon what it was to be human cannot be overestimated, in particular the opportunities for establishing value-systems based upon impersonal criteria such as money and possessions. The result has been the creation of cultures populated by people driven towards the achievement of these measurable goals, and victims for example to the *emotional* appeal of advertisers who ruthlessly exploit emotional defencelessness to sell their products. The more completely we are overwhelmed by impersonal criteria of value the less adequate will be our emotional resources to deal with the human problems which life throws up, and the more likely we will be to resort to further mechanical "solutions" such as drug-based therapies which address symptoms rather than causes, and consequently work to disguise the sources of the problems.

It is necessary to progress through five levels of truth before we arrive at a level which is capable of controlling all the other levels, and that final level is living, personal and organismic, dependent upon the capacity of organisms to adapt and relate to their environments in a flexible and creative way. Grammatical correctness, semantic clarity, referential focus, and moral precision do not exonerate themselves from falsehood; only the perception which arises from personal truth, and the participation which enables it, can assess each of them, for only personal truth is endowed with the understanding which is capable of acting as their supervenor.

Perceptions of truth are performative acts dependent upon the relationship between the perceiver and the universe. To the extent that he stands in a relationship of fundamental faithfulness to that universe, neither distorting it nor being distorted by it, those perceptions will themselves be true; to the extent that he does not stand in such a relationship his perceptions will be false. A bad tree cannot bear good fruit, nor a good tree bad fruit. We confess Jesus of Nazareth as the Christ if and only if we perceive in him that orientation to the world which involves and arises from such utter faithfulness, in which the self is first emptied of itself, and then refilled, lost and then found, laid down and then restored. In other words, in Jesus we perceive the consummation of personhood. Believing in him, dwelling in him, is our shorthand expression for striving to attain that same orientation to the world which is true understanding, and which confers upon us that capacity for discernment which Paul describes as "the mind of Christ". This "mind" is not divorced from reality, but arises within the closest possible encounter of man with reality, a relationship of indwelling in which our whole being is transformed by and conformed to the nature of the created world. The transforming vision enables us to change by providing the kind of integrative focus for our lives that more mundane goals do in the performance of daily chores. In traditional terminology, we pass through death to new life.

The concept of self-loss or self-denial which is implicit in other-centredness should not be confused with asceticism, which easily becomes a form of self-affirmation as we bask in self-righteousness at our self-denial. The logic is rather as follows: I am nothing, being a creature made from dust, and therefore I can and must be nothing to myself; God is everything, and to him all things owe their existence, even the things we fashion from the clay we find beneath our feet; but God has shown that I am not nothing to him by revealing himself to me in the life and love of his Son, and therefore because I am something to God I can and must not be nothing to myself. "Love your neighbour as yourself" therefore ceases to be perplexing as to the "yourself", for I love myself appropriately exactly insofar as I perceive myself to be a child of God whom God loves, and as such not nothing but something, and I love my neighbour appropriately exactly insofar as I love him as a child of God whom God loves. What rationalist science cannot but see as a chance concatenation of elements, and as such as dust, I also see as dust, but what I see as dust I also see as a creature, and the creature as a child, and the child as a child of the Father, a son.

The debate between reductionist and holist cannot be resolved by one or other producing evidence to establish his position, for evidence is itself part of the formalism of the science, and no formalism is ever sufficient to interpret itself. The reductionist will not cease to be a reductionist when superior evidence is presented to him, but when a superior *vision* impresses itself on his mind which transforms his understanding. The evidence will not change, but his interpretation of it will.

AUTHORITY

Wherever a Christian goes he will encounter the suspicion that he wishes to convert or convince. There cannot be logical bridges between incompatible systems, and therefore conversion cannot be a matter of *logic* or *proof*. The possibility of conversion rests upon the experience of *authority*, the recognition of that in the other which one needs to know, believe, do or become, and which moves us to change.

Response to authority is a strong theme running through the progression from childhood through adolescence to adulthood. A child accepts the authority of parents and teachers relatively co-operatively; an adolescent is characterised by rebellion against the prevailing values held by parents and society; an adult gradually reintegrates himself into a prevailing system of authority, partly by conviction and partly through necessity. Despite this adult reintegration there generally remains enough of the adolescent in each of us to maintain resentment of anyone trying to tell us what is good for us. There are echoes of our cultural rebellion against teleology, which relies upon recognition of legitimate ends for its direction.

Parents are often concerned to prevent their children making certain kinds of mistakes, particularly physically dangerous ones. They use their experience to foresee danger from fire, traffic, falls, and they both warn and restrain their children appropriately. Children need to learn to make mistakes, but it is equally necessary for them to be protected from serious harm. As we become physically more independent the focus changes to psychological and social dangers, but at the same time adolescence begins to assert itself, so that even in relatively uncontroversial activities such as learning mathematics students commonly refuse to be guided by their teachers' foresight. A short-cut in one problem avoids tackling a difficulty which must be mastered; refusal to take certain kinds of result to heart denies pupils the unconscious skills necessary to spot solutions to problems. But the admonition of the teacher is ignored: "I cannot see the point of your pedantry, so I will not accept that you are right" is an attitude we can all recognise in ourselves, and it easily changes into "nobody knows better than I do what is good for me", an attitude which has assumed almost legal dimensions in our culture. Nobody has the right to dictate what another should do, or may read or watch, as the liberally-minded argue over censorship.

Our determination not to respect authority reflects our individualism, but seems incongruous in relation to almost everything we know, which depends upon the work and authority of previous generations. Neither language nor understanding is individual: both arise within a culture (what Jung called a "collective unconscious"). Unless I accept the authority of a tradition or place myself under the guidance of a teacher or supervisor I will be forced to depend excessively upon my own resources, which are small in comparison with the collective efforts of mankind.

We accept the best of the physical creations of past generations, the formal tangible authority of fact manifested in buildings and literature, for example, but we are reluctant to accept the non-formal authority of their collective wisdom. Lacking that wisdom we find that we cannot direct our more tangible existences. We blame the tangible, formal objects around us for our ills: T.V. sets become "good" or "bad"; motor cars become "good" or "bad"; wealth and property become "good" or "bad". Yet these things are no more good or bad in themselves than a knife, which can be used to chop onions or to kill. The use governs the morality, but the use is directed by wisdom and understanding, the very thing we most neglect in our educational and cultural systems.

Mathematical authority is not present in living groups so much as in a corpus of scholarship and expertise written in books and handed down. The same is true of theology where, as in mathematics, certain insights are credited with particular importance, and made canonical: the collective authority of successive generations in the Church says, "this dogma points the way forward because it is pregnant with meaning and truth"; subsequent generations (at least until recently) respected these affirmations and built upon them.

I hope that I have gone some way in these pages to disarming the usual reply to this observation, namely that this is because mathematics deals with facts and proofs whereas theology is a matter of opinion. I have attempted to alter the reader's understanding of mathematics from an emphasis upon proof to one upon insight, from confirmation of hypotheses to their invention. Confirmation is usually the work of hacks, skilled hacks perhaps, but hacks nevertheless; the latter is the work of genuine creative thinkers pushing at the boundaries of the unknown. These intimations of fruitfulness lead us via the higher ground to vantage points from which to look upon the promised land in which we are freed from our misconceptions by the truth, which releases us from the prison of the self into the world of the other.

Education in wisdom can only ensue from the rehabilitation of authority. We have to start to learn to respect the authority of former generations' and cultures' insights by cultivating a restorative hermeneutic which will make available to us those perceptions of permanent value about the human condition such as are contained within the scriptures of the world's religions. Mathematics and theology offer us models of authority based less upon indubitability than upon genuine understanding. They are repositories of wisdom; one is highly formalised, the other less so, but both recognise as authoritative the understanding of the world which expresses itself in those formalisms.

It should be clear that this does not invite restoration of the authority of the teacher as an *individual* (there are already too many teachers motivated by the need to be thought to be authorities), but as a conveyor of the collective wisdom which is the basis of his authority. The same should be true of ministers of religion: their function is to act as conveyors of the wisdom embodied in the Church (much as a rabbi, imam or guru does in his religion). The problem is that the churches have lost their way by eschewing any such claim to authority, and in so doing they have deprived their ministers of any authority so that cults of personality are born centred upon such figures as the "eccentric parson".

EDUCATION IN CHRISTIANITY

Is it in fact true that the Christian wishes to convert and convince? Is that what preaching and proclamation involve? I do not see myself as having a responsibility to convince or convert, but to present the Christian Gospel as it is. The reason why this seems odd is that evangelicalism insists that the Church is engaged in the business of saving souls, of coaxing people into the ark of salvation by persuading them to give assent to certain formal statements of belief. But suppose this is not its function? Suppose the whole idea that by persuading people to believe in Christianity you materially aid and abet their salvation is *wrong*? What would the function of the Christian disciple be then?

It seems to me that the most neglected role of the disciple is that of *teacher*, by which I do not mean *schoolteacher*; I mean someone able to present Christianity as it is, to place it on view on the shelves in the shop of human ideas. You must know what it is if you are to choose it, and if you are to progress from being a neophyte to a disciple you will require the services of a teacher to help you along, just as you will if you are to learn the piano or mathematics. But neither music nor mathematics can be taught properly by people who are unconvinced of their value. I teach mathematics because I believe that by understanding mathematics better my students will obtain a perspective on dimensions of existence which would otherwise be closed to them, and will acquire skills which will serve them well in a vast range of activities, in other words because I believe that it will enrich them as people by opening their eyes. They may not share my enthusiasm; they may prefer history; but at least they are being given the chance to enthuse, and somewhere they may acquire from their education a fascination which will accompany them through the rest of their lives. Most religious education is provided by teachers who are far from enthusiastic about it, and examination syllabuses often seem to defy candidates to exhibit any religious conviction.

We make our choices by reference to what we want to do. To send rockets to the moon, study sub-atomic physics, investigate numbers, or build bridges we will select a part of mathematics. If we wish to be edified, amused, secure, open-minded, closed-minded and so on we will choose a part of religious experience accordingly. In all such cases we make fundamental choices about the degree of reality afforded us by what we study, and we are inevitably shaped as a consequence. Let us not think, "here is mathematics, certain and true; there is theology, uncertain and unproven", but "here is mathematics, which says something about the way the universe is, and here is theology, which says something about how life is". Einstein's theories of relativity are two among many cosmologies; how am I to decide? Behold and see! Christianity is one among many religions. How am I to decide? Behold the man!

When St Paul in 1 Corinthians 2:21 speaks of having "the mind of Christ" he is referrring to that state which is characterised by spiritual discernment, the discernment which penetrates to the truth of things and enables us to be in relation to things in a way possible only with the help of the Spirit of God. He does not imply any usurpation of the place of Christ, still less that we in any way become his equivalent in wisdom, goodness or stature; his purpose is to emphasise that singular orientation toward the world which recognises the utter centrality of the being of God. The translation involves a movement of understanding from the appearances of things or their formalisation to the meaning and truth which can only be discerned by dwelling in them, by their becoming the subsidiaries through which we perceive the truth as their focus. It involves acknowledging and resolving the paradox of self-centredness which pretends to be objective and based upon impersonal foundations into the integrated personhood which acknowledges the inevitable personal and social coefficients in knowing and being but which nevertheless finds its centre in the other. It involves being without foundations, being sustained by the living relationship with the other which shapes our minds into the pattern of the mind of Christ.

The reorientation achieved by this inversion of our understanding of Christianity is the equivalent of a Copernican revolution, for it turns the nature of Christian activity inside out and leads the way to a richer, more profound integration of Christian doctrine based not upon our need to convert, but on our mission to proclaim. The function of Christian doctrine ceases to be to supply the formal statements of belief necessary if people are to be saved, and becomes the means whereby we are led to a deeper vision of the Truth and a greater strength to honour the responsibilities which arise for us from that perception.

The theologian to whom we owe this Copernican revolution in theology is Karl Barth, who reaffirmed the centrality of the "wholly other" rather than the self. His work burst like a new star upon the darkness of liberal theology, until scepticism and self-centred reason regrouped their forces. His treatment bears striking resemblance to the treatment of Galileo, and the issues are curiously similar: do we find our centre in God, or does God find his centre in us; does the earth orbit the Sun, or the Sun the earth? Where is the centre of all things, the Truth which embraces all things and makes them whole, is it in the self or in the Other?

In the introduction I gave some indications of the means the churches have to employ to decide upon doctrinal agreement, and I identified one or two contemporary issues which are fiercely debated such as the Virgin Birth and the historicity of the Empty Tomb. The relocation of our centre and the reorientation of our minds that I have argued for makes it possible to give a brief response to these questions as a concrete example of one way in which the aims of this book find themselves confirmed in practice.

The question we must address is why we might wish to affirm the truth of the Virgin Birth or the Empty Tomb. I have already, in the chapter on truth, given a partial answer to such questions. The most important reason belongs to the second book I have in mind rather than this one, but it may be given in summary form here. If beliefs and doctrines are conceived as sets of statements which we must affirm in order to secure salvation, as is widely held and taught, and if certain religious rituals such as worship, Baptism, the Eucharist, are regarded as obligatory if we are to be deemed worthy of everlasting life, as is equally widely held and taught, then the formal has usurped for itself the place of the non-formal, the Law has again assumed ascendancy over the Spirit, and Christ died in vain. But if Christianity is about the proclamation of salvation rather than its achievement everything suddenly appears in a new light. Beliefs cease to be a burden, things to be affirmed despite the fact that we do not believe them (just to keep on the right side of God), and become vehicles for a deeper perception of the Truth. Instead of saying or feeling "do I really have to believe that?", we can begin to recognise that our inability to believe it may mean only that we are denied access to certain kinds of insight that come through it. The stance I have assumed throughout this book is that the claims of the past should not be abandoned because they do not fit our presuppositions, but be permitted to challenge those presuppositions. I would wish to affirm the truth of the Virgin Birth and the Empty Tomb, not because I would not believe that you were saved if you found it impossible to believe in them, but because I am persuaded of their truth by their richness; their scandal drags me, screaming with self-centred indignation, into the very centre of the being of God.

To clarify this point further I give an example where I believe myself to be impoverished by my inability to appreciate something. I have in mind not a religious instance, but a literary one. Many of my friends have a great love for Jane Austen's novels. Hearing them speak one encounters echoes of a plagiarised Johnson saying "when you are tired of Jane Austen you are tired of life". Unfortunately, try as I will, I find myself completely unable either to like Jane Austen or to read more than the first few pages of any of her books. By analogy with the cases cited above I do not doubt for a moment the truth of the valuation my friends make of Jane Austen, and I do not for a moment set myself against their valuation of her insight or importance, and I do not deny that her insight into the human character is extremely amusing and profound. I simply have to confess that I cannot bring myself to read her, accept the approbation of some that that will earn me, and acknowledge that I am very much the poorer for it. I recognise the sadness that my block about her causes some of my friends, and their genuine puzzlement that someone with whom they have so much in common in other fields can be such a "Philistine" in this one, and the limitations on our communication my blind-spot imposes. But I regard those who find themselves unable to benefit from belief in certain doctrines in the same way (but from the perspective of my "friends"). Notice that I am in no sense impugning the truth of the value placed upon Jane Austen in the one case, and in no sense surrendering the truth of the doctrines in the other.

At its sharpest this distinction obtains for those who reject the Christian faith altogether, or who embrace it in a "self-centred" way. What they lose is not salvation, but the faith and hope and love which arise from having access to the Truth at the heart of all things. They are simply and profoundly the poorer for their unbelief and error. But that is always the way with truth: to lie, or to be in error, or to refuse to acknowledge the truth does not lessen the truth; it lessens our selves. It is this perception which is so characteristic of Jesus' whole life. We have no instances recorded of his anger directed against unbelievers, but we know in some detail both of his anger at those who misled people by perverting religion (the polemics against the "hypocritical" Jewish leaders), and those who reduced religion to a matter of commerce (the cleansing of the temple). In fact we have clear confirmation of our view in the story Luke tells about Jesus arriving near Jerusalem: "O Jerusalem, Jerusalem, killing the prophets and stoning those who are sent to you! How often would I have gathered your children together as a hen gathers her brood under her wings, and you would not!" (Luke 13:34). The great sin for Jesus is not unbelief — he has both compassion and patience for unbelievers — or even moral sin — he has both compassion and forgiveness for sinners in this sense also — but sin against the truth. For someone to perceive the truth, and to pretend that he is not responsible for the truth perceived, is sin, and to perceive the truth and then to turn away from it or to pervert it for his own gain is the greatest sin of all. "Woe to you, scribes and Pharisees, hypocrites! because you shut the kingdom of heaven against men; for you neither enter yourselves, nor allow those who would enter to go in. Woe to you, scribes and Pharisees, hypocrites! for you traverse sea and land to make a single proselyte, and when he becomes a proselyte, you make him twice as much a child of hell as yourselves." (Matthew 23:13–15.)

What is the Mind of Christ? Jesus perceived, and what he perceived invested him with responsibility, the responsibility to share his vision no matter what the personal cost. He not only taught that when we find a pearl of great price we should surrender everything in order to secure it; he lived by that principle. And the pearl is the Truth of the Word that proceeds from the mouth of God. To betray the Truth which is perceived in the Other is to die a greater death than the death of the body; it is to die the death of the soul; it is to cease to be. Jesus did not preach a gospel purporting to be a panacea for all ills. Unlike many modern evangelists, he did not preach that if you turn to him all will be well in the sense that your health will improve, your career be more successful, your mortgage arrears disappear, your ulcers vanish, or your life be prolonged. He came with empty hands prepared to call and make disciples of all those weak enough and strong enough to bear the burden of responsibility that perception of the truth which lies in God entails. The prophets before him saw how great that burden could be, and many begged to be released from its requirements; but his yoke is easy, and his burden is light.

As a full-time clergyman I have walked round hospitals talking to the sick, I have walked round factories talking to the healthy, I have walked round towns talking to the irreligious, and I have walked round parishes talking to the religious. In all four cases I was struck by the fact that I had nothing to offer any of them which would make sense to them in their terms (that is, in terms of a set of needs and reasons and values still centred upon themselves). I was also conscious that many of my colleagues resolved the tension between the true answers provided by the Gospel and the false questions posed by the world by converting what they had to say to conform to what their hearers wanted to hear, and having begun with high hopes of resisting the pressure to follow I eventually found myself doing the same. But this is to anticipate what I hope to say in the other book I have mentioned. This book arose from an attempt to think through the issues raised by these experiences, and from the insight that we have two problems to solve: the problem of the overwhelming formalism and self-centredness of human reason, which rules Christian teaching out of court without a hearing; and the problem of the image of the Church, which rules Christian ministry out of court without a hearing.

The personal encounter of a man with Christ or the encounter of a scientist with the universe is an encounter in love, trust and obligation wherein he is called, and perceives the call, to relinquish the freedom of the subjective person to believe as he likes, and the freedom of the objectivist to refuse to believe at all, in favour of the freedom of the responsible person to believe as he must. His response will be to share his vision with others, a vision which he must describe despite its indescribability, a vision which he will seek to brighten by learning from similar descriptions of others, despite their inevitable inadequacy. Sharing in the community of faith which sets and confirms the standards by which we dare attempt to formalise that which lies beyond all formalism, the Christian, be he scientist, clergyman, philosopher or layman, relies ultimately on the one person of Jesus Christ. The way which is disclosed to us when we pursue his Way, and the truth which is realised when we participate in his Truth, give rise to the life which he made possible for us by the way he lived his Life. The finitude and the limitation which ensure that we will always fall short of that goal must not lead us to espouse lesser goals, however often we stumble. Nor should it, in the interests of a less provocative or more accessible Christianity, lead us to dilute the awesome implications of the other words associated with "I am the Way, the Truth, and the Life: no one comes to the Father, but by me".

INDEX

Absolutism, 67f
Anselm, 22, 41, 132, 139
Art, 100
Authority, 173, 181, 198
Axioms, 15f, 41f, 105

Barth, Karl, 14, 204
Bonhoeffer, D., 91

Carnes, J., 15f
Certainty, 46, 67f
Change, 8, 119, 180
Church, 9, 11, 39
Circularity, 67
Clerk Maxwell, J., 22
Concepts, 32, 140, 167
Conflict, 110
Conversion, 67, 74
Conviviality, 10, 36
Creativity, 33ff

Death of Jesus, 116
Democratic Fallacy, 13, 26, 35
Dual Control, 134

Education, 74, 201
Error, 110, 121f
Eucharist, 104
Evolution, 175

Faithfulness, 88
Falsehood, 144
Feedforward systems, 102f
Formalisation, 20, 60, 146, 158
Formal system, 9, 31, 60f, 170
Foundations, status of, xix
Frege, G., 78f
Fundamentalism, xxi, 187

Gädel, K., 54, 81, 85, 128f, 175

Immunisation strategy, 90
Indwelling, 72, 130

Language, 99, 130
  and formal systems, 9 and *passim*
Logic, 41–59, 114, 128–130
Love, 93, 102

MacIntyre, Alasdair, xix, 24
Marriage, 22, 102
Measurement, 96f
Mechanised Intuition, 77
Mental Image, 99
Ministry, xxi, 207
Music, 4, 32, 141

Newbigin, Lesslie, 13
Newtonian Mechanics, 93f

Participative Reason, 115
Perfectionism/Inversion, 3, 29, 97, 121, 135f, 147
Play, 119
Polanyi, Michael, xvii, 3, 27, 30, 35, 66, 88, 121, 130, 179
Positivism, 45
Post-Critical Philosophy, 64–67
Protection Racket Principle, 97, 156

Questioning, 85
Quotations, 133, 168

Reality, 23, 65, 82, 99
  mathematical, 81, 95
Reason,
  and Emotion, 1, 102, 110
  other-centred, 89f
  self-centred, 89f, 136

Relativism, 68, 71
Religion, 46, 90, 127
   varieties of, 160

Scientific Community, 10f
Scientism, 28
Subjectivism, 108
Supervening Vision, 73, 191

Teleology, 37

Theological Imperative, 88, 109, 127, 156, 165
Things, classification of, 143
Time, 162f
Torrance, T.F., 21, 65
Tragedy, 162
Truth, classification of, 132, 137, 140
   historical, 150

GENERAL THEOLOGICAL SEMINARY
NEW YORK

DATE DUE

HIGHSMITH #45230